PEREGRINE BOOKS

GEOMETRY AND THE LIBERAL ARTS

For an architect like Vitruvius and artists like Leonardo and Dürer the fascination of geometry lay in its contribution to solving problems of order, proportion and perspective. With more recent excitement over the 'new mathematics' we have tended to neglect geometry's lasting appeal to artists, scientists and philosophers; but in this book Professor Pedoe, taking these three great practitioners as his principal examples, redresses the balance and traces the effect that geometry has had on artistic achievement.

After a refreshing look at Euclid's *Elements* (still the inspiration of modern geometry) he guides the reader through the discoveries stimulated by the theory of perspective and discusses form in architecture, projective geometry, and mathematical curves as well as the concept of space.

Professor Pedoe, whose infectious enthusiasm dominates a fascinating and informative book, also includes suggestions for practical and theoretical exercises in construction, 'curve-stitching' and making geometrical models.

Dan Pedoe, B.Sc. (Lond.), B.A., Ph.D. (Cantab.) is Professor of Mathematics at the University of Minnesota. He was born in London and is now an American citizen. Among his publications are *Methods of Algebraic Geometry,* written in collaboration with Sir William Hodge, and *The Gentle Art of Mathematics* (Penguin 1963)

GEOMETRY
AND THE
LIBERAL ARTS

DAN PEDOE

PENGUIN BOOKS

Penguin Books Ltd, Harmondsworth, Middlesex, England
Penguin Books Inc., 7110 Ambassador Road, Baltimore, Maryland 21207, U.S.A.
Penguin Books Australia Ltd, Ringwood, Victoria, Australia
Penguin Books Canada Ltd, 41 Steelcase Road West, Markham, Ontario, Canada
Penguin Books (N.Z.) Ltd, 182–190 Wairau Road, Auckland 10, New Zealand

—

First published 1976

—

—

Made and printed in Great Britain by
Hazell Watson & Viney Ltd, Aylesbury, Bucks
Set in Monotype Times

To Judith Mary,
and all creators of
beauty

CONTENTS

LIST OF PLATES

PREFACE

GEOMETRY has always been one of the seven liberal arts, which combined the *trivium*: grammar, rhetoric and dialectic, and the *quadrivium*: arithmetic, geometry, astronomy and music. A book, originally written by the great Cassiodorus in the sixth century as a kind of encyclopedia of profane learning which he thought his monks needed to enable them to understand the Bible, contains a brief exposition of the seven arts. This book was widely used in the Middle Ages by laymen as well as priests. The number *seven* has always had magical associations, and Cassiodorus justified his choice of seven essential subjects with the scriptural quotation: *Wisdom hath builded her house; she hath hewn out seven pillars*. Of course the various arts did not remain static, and the Renaissance witnessed a great development of the quadrivium. Arithmetic and geometry, with the addition of algebra, became *mathematics* in the course of time, but it can be said that until a short time ago *geometry* was regarded as an essential part of any liberal education; that is of any education not definitely oriented towards a profession.

Geometry has enormous historical interest, and also practical and aesthetic appeal. This book can be taken by the general reader as a diversion into the by-ways of history, with glimpses of the enormous importance of geometry to such people as Vitruvius, Albrecht Dürer, Leonardo da Vinci and Thomas Hobbes, to name but a few of the characters who appear in these pages. On the other hand, Exercises are given at the end of each chapter for those who would wish to experience the aesthetic appeal of geometry by carrying out simple constructions, inventing patterns, drawing and stitching curves and envelopes, and finally venturing into the exciting domain of *mathematical proof*.

Very little previous acquaintance with geometry and algebra is assumed, but there are places where statements are made which can only be verified by readers with some training in mathematics. The treatment is certainly not encyclopedic, and there are many omissions. In writing this book I have naturally followed my own preferences in reading, and in discussing those aspects of geometry which I think important. This is therefore a very personal book.

Chapter 1 is a fairly slow introduction to those aspects of geometry which can be found in Vitruvius, and the reader is encouraged to use ruler and compasses, perhaps for the first time. Chapter 2 reveals the interest and excitement I discovered in the works of Albrecht Dürer, which I hope the reader may share. The theory of perspective stimulated the development of geometry, and is given much attention in this and later chapters. Chapter 3, on Leonardo da Vinci, completes a fairly detailed study of three great historical characters and their attitude towards geometry.

Chapter 4 discusses Form in Architecture, and the various theories which have been fashionable from time to time. In Chapter 5 an introduction to Euclid's great *Elements of Geometry* is given, the medium being a far lesser work, the *Optics*, but one which indicates the possibilities of *proof* in geometry. Chapter 6 deals with Euclid proper, the book which once shared the honour with the Bible of being the world's greatest best-seller.

Chapter 7 introduces both Cartesian and Projective Geometry, and shows how some of the great theorems of projective geometry can be used in perspective drawing.

Chapter 8 discusses some of the lovely curves which geometers should certainly not keep to themselves, from conic sections to equiangular spirals. The reader is shown how to produce these curves in a variety of ways. The focal properties of the conic sections are displayed, but if the proofs which are given seem to be involved, the reader can still enjoy the methods given for drawing these sections of a cone. It is in this chapter that mathematical embroidery is introduced.

The final chapter deals with *space*, from a number of aspects, with adventures in two, three and four dimensions.

I have enjoyed writing this book, and in doing so have come across many attractive geometrical theorems previously unknown to me and, I am sure, to other lovers of geometry. Experience with a wide variety of audiences, in many different countries, has shown that the topics discussed here interest the majority of literate people, and that *the pleasure principle*, which is actively pursued in these pages, should be considered of fundamental importance in all education, and especially in mathematical education.

A list of the books consulted during the preparation of this book is given on p. 287. Since it was evidently impossible to return to original sources in many cases, I have made extensive use, with due acknowledgements, of material in some of the works listed. However, the interpretation given to geometrical references is my own and differs, in some cases, from that of so-called authorities. Historical references in the text vary in detail, but dates, when these are known, are assigned to the *dramatis personae* of these pages.

As an alumnus of both Universities, I especially wish to thank both the Cambridge University Press and the Princeton University Press for permission to use and to quote from published material. I also wish to acknowledge a grant from the Center for Curriculum Development of the University of Minnesota, which enabled me to start writing this book in the summer of 1971. Finally, I wish to thank my publishers, although publication has been long delayed, for producing a very attractive book.

School of Mathematics, DAN PEDOE
University of Minnesota,
Minneapolis, Minn. 55455, *U.S.A.*

·1·

VITRUVIUS

THE word *Geometry* is of Greek origin, and signifies *land-measurement*. There is an often-quoted passage in the writings of the Greek historian Herodotus (485–425 B.C.) on the origins of geometry.

They said also that this king [Sesostris] divided the land among all Egyptians so as to give each one a quadrangle of equal size and to draw from each his revenues, by imposing a tax to be levied yearly. But every one from whose part the river tore away anything, had to go to him and notify what had happened. He then sent the overseers, who had to measure out by how much the land had become smaller, in order that the owner might pay on what was left, in proportion to the entire tax imposed. In this way, it appears to me, geometry originated, which passed thence to Hellas.

This is a plausible explanation, and we accept it for this reason, and also because no other explanation is available. Until we reach fairly modern times, roughly the last seven hundred years, our knowledge of the past is mostly derived from translations made by European scholars from translations made by Arabian scholars of Greek and Hindu manuscripts. There are enormous gaps in our knowledge, since many works did not survive to be translated into Arabic, and those which did survive were translated, amended and transformed, quite often, into works which bear little resemblance to the originals. As a result we are often very uncertain about the original meaning of words and phrases used by Greeks and Romans in their writings. We shall return to these matters from time to time. It suffices for the moment that we do have some explanation of the origins of geometry, and of its early economic importance.

Practical geometry was certainly practised by all ancient peoples. Building huts, excavating caves, erecting tents, all depend on a sound intuitive notion of geometry. At a higher level of sophistication, measurement of distances in terms of a unit length, the idea of a right angle, the evaluation of the area of a rectangle, all this is geometry. But the ancient Greeks went far beyond what was necessary to establish rules for practical geometry, and in doing so set their seal on the development of European civilization as we know it today.

Our discussion will not be chronological and we begin with the Roman architect Vitruvius who lived several hundred years after Euclid and seems to belong to the early Christian period, although his exact date is uncertain. He effusively dedicated his *Ten Books of Architecture* to the reigning Emperor who is thought to have been Augustus. The Books were written to advise the Emperor on the construction of buildings and on everything concerned with architecture.

An architect, says Vitruvius, should be skilful with his pencil, instructed in geometry, conversant with history, should have followed the philosophers with attention, should understand music, have some knowledge of medicine, be interested in the opinions of jurists, and acquainted with astronomy and the theory of the heavens. He reveals his own knowledge in these areas in his ten books and shows why he thinks them important to an architect.

He justifies some knowledge of history by remarking that an architect should be able to explain what he does to inquirers. For example if an architect used Caryatides (Plate 1 is an illustration of these from the first French translation of Vitruvius in 1684) he should be able to explain their origin. Vitruvius explains that they can be traced back to the unhappy state of Caryae. This was a state in the Peloponnesus which sided with the Persian enemies of

Greece. The Greeks having won their war, they took the town (many Greek states were single cities at that time), killed the men, abandoned the state to desolation and carried off the wives into slavery, without permitting them, however, to take off the long robes and other marks of their rank as married women. They were obliged not only to march in the triumphant procession which followed the victory dressed in this manner, but to appear forever after as a type of slave, burdened with their shame, to atone for their State. The architects of the time designed public buildings with statues of these women, apparently carrying the weight of the buildings, so that the sin and punishment of the people of Caryae might be known and handed down to posterity . . .

Vitruvius says that a knowledge of philosophy (which included what we call *physics* nowadays) makes an architect high-minded and unpretentious, courteous, and just and honest without avariciousness . . . The reasons that Vitruvius gives for an acquaintance with music are more practical. An architect would have to be able to design war-machines, such as ballistae and catapults. These depended on stretched strings, and according to Vitruvius, if the stretched strings were not in tune, the course of the projectile would not be straight. For this simple tuning rule to work, the architect would have to make sure that all the strings were of the same length, and of equal density, and so the tensions would be equal, but Vitruvius omits these details.

Another connection between the profession of architect and music arose in the design of theatres. In order to increase the power of the actor's voice, bronze vessels were placed in niches under the seats, and the architect had to design them to resonate in the correct way. He also had to design the elaborate Roman water organs and these obviously required a knowledge of music.

The need for an architect to have some knowledge of medicine, to be interested in the opinions of jurists, and to be acquainted with astronomy and the theory of the heavens, will be justified as we survey Vitruvius' concern with the whole subject of architecture.

Architecture, says Vitruvius, depends on order, arrangement, eurythmy, propriety, symmetry and economy. Here we have a number of words which have quite different meanings nowadays. *Order*, says Vitruvius, gives due measure to the members of a work considered separately, and *symmetry* gives agreement to the proportions of the whole. It is an adjustment according to quantity. 'By this', Vitruvius explains, 'I mean the selection of modules from the members of the work itself, and constructing the whole work to correspond.' *Eurythmy* is beauty and fitness in the adjustment of the members of a work. Symmetry is a proper agreement between the members of the work itself, and relation between the different parts of the whole general scheme, in accordance with a certain part selected as standard. 'Thus,' Vitruvius goes on, 'in the human body there is a kind of symmetrical harmony between forearm, foot, palm, finger and other small parts. So with perfect buildings.'

This last statement could easily be the basis of a long dissertation! The artist, seer and poet William Blake (1757–1827) was probably fairly near to Vitruvius in his use of the word symmetry in the poem:

> Tyger! Tyger! burning bright,
> In the forests of the night,
> What immortal hand or eye
> Could frame thy fearful symmetry?

We still talk of the two halves of a human face never being symmetrical, and the idea of *balance* is implied in this usage. In mathematics, symmetry has a very definite sense. If there is a one-to-one mapping between the points of a set S which maps S onto itself, then S is said to be *symmetrical* with respect to this mapping. We shall say more about this in Chapter 9. It was in selecting parts of the human body as a standard module, if Vitruvius was really the first to do this, that the first links were forged in an historic chain which ties together Vitruvius, Albrecht

Dürer, Leonardo da Vinci, countless other artists and the modern architect, Le Corbusier, who is discussed later.

Geometry first enters into the Ten Books when Vitruvius occupies himself with winds. He was very much concerned with health and comfort, and wanted the directions of streets in a city to be planned so that certain winds did not howl down the streets. He mentions eight major winds, and sixteen minor ones. The major ones are called: Septentrio, Aquilo, Solanus, Eurus, Auster, Africus, Favonius and Caurus, and each had its own distinguishing features and blew at certain times of the day. The minor ones are also shown in Figure 1. (We must interpolate the remark that

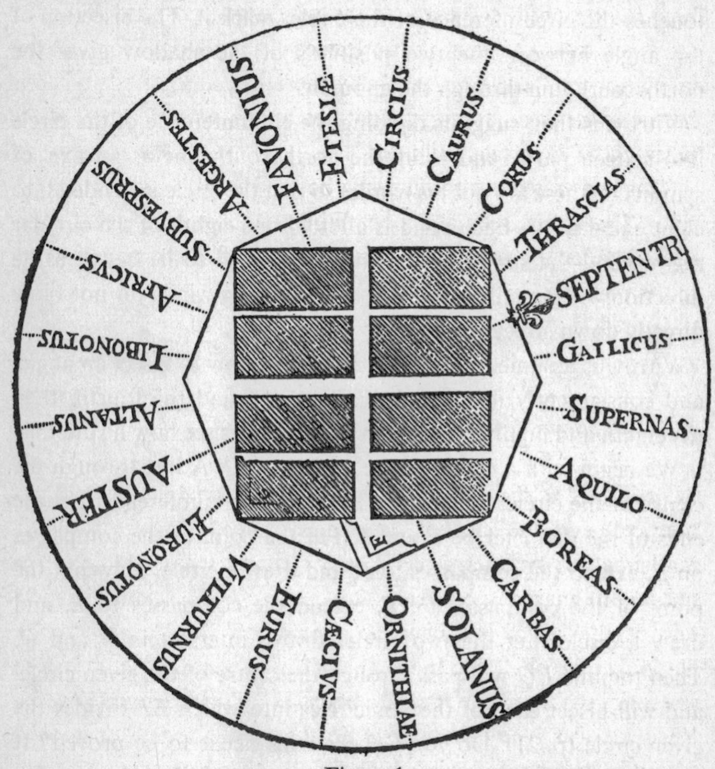

Figure 1

there was great controversy amongst the ancients regarding the number and names of winds. Aristotle, Seneca, Pliny and others all disagreed with each other, and not one agrees with Vitruvius.) To plan the streets of a city, the first thing to do is to find the north–south line.

In the centre of the planned city, on a level place, set up a bronze gnomon, or shadow-tracker. At about the fifth hour of the morning, mark the end of the shadow. Then, opening your compasses from the gnomon to the point which marks the length of the shadow, describe a circle. In the afternoon watch the shadow of the gnomon as it lengthens again, and when it once more touches the circumference of the circle, mark it. The bisection of the angle between the two positions of the shadow gives the north–south line through the gnomon.

Vitruvius then suggests dividing the circumference of the circle into sixteen parts, and using the north–south line as an axis of symmetry (these are not his words) so that the circle is divided into eight equal parts. Each wind is allotted one eighth of the circular region, and the streets of the city are aligned to lie between the directions of prevailing winds, so that strong winds do not blow directly down streets and alleys.

Vitruvius assumes that his readers know how to bisect an angle, and consequently, by further bisection, to find the fourth, then the eighth and finally the sixteenth part. Let us see how he did this.

We begin with a circle, centre A (Figure 2). A line through the centre of the circle, a diameter, bisects the circumference. Let the ends of the diameter be E and F. Put the point of the compasses on E, extend the compasses to F, and draw a circle. Now put the point of the compasses on F, extend the compasses to E, and draw a circle. Let the two circles drawn intersect in P and Q. Then the line PQ will pass through the centre of the given circle, and will bisect each of the semicircles into which EF divides the given circle (p. 21). Do you feel that this needs to be proved? If you do, what do you mean by *proof*? (See p. 145.) For the moment the construction can be accepted as a correct one.

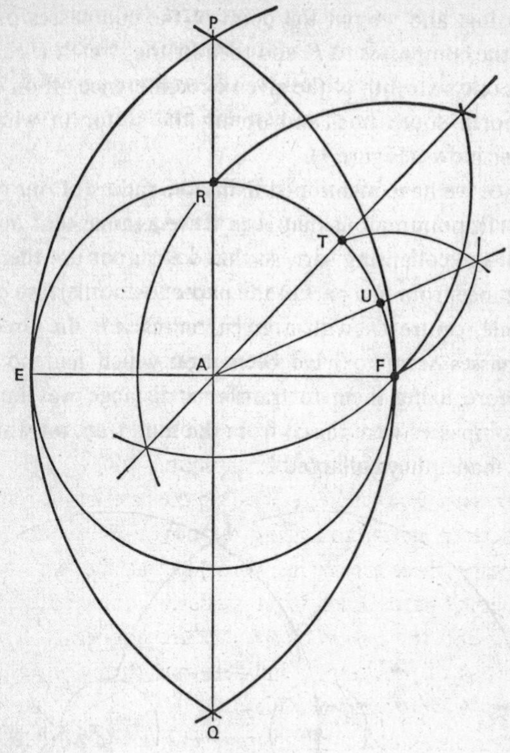

Figure 2

If the semicircle on EF as diameter and centre A, is bisected at R, we now want to bisect the arc RF. We carry out a similar construction to our previous one. Point of the compasses on R, extend to F, draw a circle with radius RF. Point of the compasses on F, extend to R, draw a circle. The line joining the points of intersection of these two circles will bisect the arc RF. If this mid-point is called T, the arc TF is one eighth of the circumference of the given circle, centre A. If we bisect the arc TF, we have our sixteenth, and if the point of bisection be U, the arc UF or the arc UT is one sixteenth of the circumference of the given circle.

Now suppose that we can fix the stretch of the compasses we

are using, and we put the point of the compasses on U, and extend the compasses to F, and then fix the stretch UF. We can then mark one sixteenth of the given circumference off on either side of the north–south line, and obtain the sector in which the wind Auster blows (Figure 1).

Since we have mentioned fixing the radius of our compasses, it is worth pointing out that even if we assume that our compasses are of the collapsing sort, so that we cannot use them to transfer a distance from one part of the paper to another, we can still draw a circle, centre O, with a given radius AB. In Greek geometry compasses seem to have been used which had no hinges, and therefore fixing them to transfer a distance was impossible. As the compasses were raised from the sand tray, on which drawings were made, they collapsed.

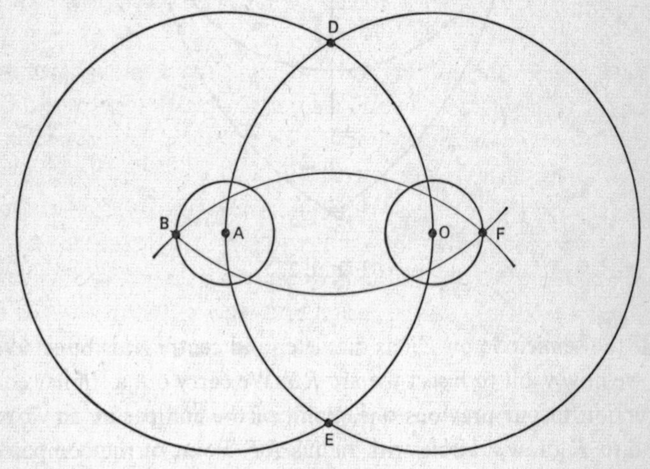

Figure 3

Let the points O, A and B be given (Figure 3). The problem is to draw a circle, centre O, with radius AB, without using the compasses (or any other instrument) to *transfer* a distance. We begin by drawing a circle centre O and radius OA, and another circle, centre A and radius AO. Let these circles intersect in D and E. Draw the circles centre D and radius DB and centre E, radius EB to intersect again in F. Now draw the circle centre O and radius OF. This is the required circle, centre O and radius equal to AB. In other words, $AB = OF$.

If we draw the circle centre A and radius AB, the modern notion of symmetry leads to a proof that this construction is correct, because the whole figure is symmetric* about the line DE. An equivalent proof, possibly easier to understand, is suggested in the Exercises at the end of this chapter (Exercise 7).

The reason for our discussion of the collapsing compasses is that in Euclid's *Elements*, the great book on geometry we shall be exploring in Chapter 6, p. 140, the fixed (or *modern*) compasses are never used to transfer distances, or arcs on a circle. What we have just proved is that the modern compasses are not superior to those used by Euclid.

Vitruvius devotes a significant amount of space, though perhaps not quite as much as Leonardo da Vinci in his Notebooks, to the proportions of the human body. This subject is introduced in a further discussion on symmetry. Vitruvius says this arises from *proportion*, a correspondence among the measures of the members of an entire work and of the whole to a certain part selected as the standard, the module. Without symmetry and proportion there can be no principles in the design of a temple, that is there is no precise relation between its members, as in the case of a well-shaped man.

We give Vitruvius' proportions. He says that the human body is so designed by nature that the face, from the chin to the top of the forehead and the lowest roots of the hair, is a tenth part of the whole height; the open hand from the wrist to the tip of the middle finger is just the same; the head from the chin to the crown is an eighth, and with the neck and a shoulder from the top of the

*Reproduce Figure 3 on a sheet of paper, and fold the paper along the line DE. What do you notice?

breast to the lowest roots of the hair is a sixth; from the middle of the breast to the summit of the crown is a fourth.

If we take the length of the face itself as our unit, the distance from the bottom of the chin to the under side of the nostrils is one third of the face; the nose from the under side of the nostrils to a line between the eyebrows is the same; from there to the lowest roots of the hair is also a third, taking in the forehead.

The length of the human foot is one sixth of the height of the body, the length of the forearm is one fourth, and the width of the breast is also one fourth. The other members of the body also have their symmetrical proportions, and it was by employing them that the famous sculptors and painters of antiquity attained great and endless renown . . .

It is the next passage in Vitruvius that has given rise to many drawings:

. . . then again, in the human body the central point is naturally the navel. For if a man be placed flat on his back, with his hands and feet extended, and a pair of compasses is centred at his navel, the fingers and toes of his two hands and feet will touch the circumference of a circle described therefrom. And just as the human body yields a circular outline, so too a square figure may be found from it. For if we measure the distance from the soles of the feet to the top of the head, and then apply the measure to the outstretched arms, the breadth will be found to be the same as the height, as in the case of plane surfaces which are perfectly square.

It will be noticed that a censor has been at work on the plate reproduced from the 1684 edition of Vitruvius. The scales at the side give the Greek, Roman and Royal (that is the French) units of comparative measurement (Plate 2). (Later Leonardo illustrated the Vitruvius concepts of human measurement in a drawing shown in Plate 3.)

To turn to another aspect of the Vitruvius notion, it has been suggested that the Romans actually crucified people in the capital X position suggested by the figure inscribed in a circle, but that early painters of the most eventful crucifixion in history thought the posture an indelicate one, and changed the position of the cross to the one we all know from countless paintings and sculptures.

It is inevitable that in a work of this sort we must be able to traverse time very rapidly, both backwards and forwards, in pursuit of an idea. We have already mentioned Le Corbusier as a modern architect who worked from *modules*. In one of his books he tells the following story: he had been working for some time with a module based on the height of 1·75 metres for the body of a man, presumably a Frenchman, and certain numbers in two scales refused to come out nicely. (We shall give more details of these calculations in Chapter 4.) Then one of his collaborators said: 'Isn't the height we are working with rather a French height? Have you ever noticed that in English detective novels the good-looking men, such as the policemen, are always six feet tall?' Le Corbusier continues: 'We tried to apply this standard: 6 feet = 182·88 cm. To our delight, the graduations of a new "Modulor", based on a man six feet high translated themselves before our eyes into round figures in feet and inches.' He used modules based not only on the height of a man six feet tall, but also on such a man with one arm stretched above his head, and in fact this last symbol was more or less taken over by Le Corbusier as his coat of arms.

Measurements, as Vitruvius points out, have from time immemorial been based on the human figure. The *foot* is an obvious example, but some of the others, such as the *palm*, the *finger* and the *cubit* have become archaic.

Vitruvius refers fairly often to the Pythagoreans, the disciples of Pythagoras. Pythagoras is thought to have lived from 572 to 490 B.C., and the Pythagorean brotherhood flourished until A.D. 415. Vitruvius writes in fairly short sections, and bases this action on Pythagorean doctrines. He asserts that the Pythagoreans thought it proper to employ *the principles of the cube* in composing books on their doctrines. They determined that the cube consists of 216 lines, and held that there should be no more than three cubes in any one treatise. If the cube is thrown down, it stands firm and steady on any side, like the dice which players throw on a board. The Pythagoreans appear to have drawn their analogy from the cube, because the number of lines mentioned will be firmly and steadily fixed in the memory when they have settled down, like a cube, upon a man's understanding . . .

We notice that 216 = 6.6.6, but this information does not tell us how the Pythagoreans associated this number with a cube. The notion that the essence of the universe can be discovered from the properties of geometrical objects, which characterizes Pythagorean thinking, has not died out by any means, and will be referred to several times in this book. In fact, Plato's remark: 'God geometrizes' was often quoted in the 1930s, when theoretical physicists were busy making models of the atom, and geometry has always been associated with rational beings.

Vitruvius tells a number of stories, and one of them concerns the philosopher Aristippus, who was shipwrecked on the coast of Rhodes. The philosopher saw some geometrical figures drawn in the sand, and cried out to his companions: 'Let us be of good cheer, for I see traces of man!' Edmund Halley's Latin edition of the *Conic Sections of Apollonius*, published in 1710, has a plate illustrating this episode. While Aristippus blissfully contemplates a pair of conjugate hyperbolas in the sand, his companions are drowning. As we know nowadays, not all men who can do geometry, if they try, are good men. Mistakes can also be made in observing alleged geometrical activities.

It was thought for a long time that there are canals on Mars, and it was a rational deduction that Mars must be inhabited. But as telescopes became better, the alleged canals were no longer seen as canals, and the idea, except for science fiction, was abandoned. We now even have close-up photographs of Mars. But attempts have been made to send coded messages to outer space which would be intelligible to rational beings of a high order of intelligence. If we wished to simulate the Martian canals, and to draw a geometrical figure on an enormous scale which would surely convince observers in outer space of the existence of rational beings on the planet Earth, we could hardly do better than to draw the famous figure given in Euclid to prove Pythagoras' great theorem: the square on the hypotenuse of a right-angled triangle is equal to the sum of the squares on the other two sides (Figure 4).

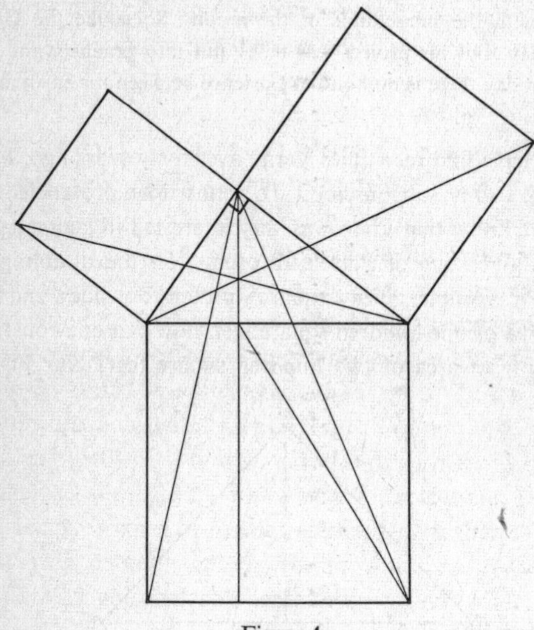

Figure 4

Euclid's proof is a logical deduction from preceding theorems on areas, which accounts for the various lines in the figure. Other proofs are possible, and one is suggested in an Exercise at the end of this chapter.

Jules Verne had the same idea in his novel *From the Earth to the Moon*, and the passage is worth quoting:

... a few days ago a German geometrician proposed to send a scientific expedition to the steppes of Siberia. There, on those vast plains, they were to describe enormous geometric figures, drawn in characters of reflecting luminosity, among which was the proposition regarding 'the square of the hypotenuse', commonly called the 'Ass's Bridge' by the French. 'Every intelligent being,' said the geometrician, 'must understand the scientific meaning of that figure. The Selenites, do they exist, will respond by a similar figure; and, a communication being thus once established, it will be easy to form an alphabet which shall enable us to converse with the inhabitants of the moon.' So spoke the German geometrician, but his project was never put into practice, and up to the present day there is no bond in existence between the earth and her satellite.

I must have first read Jules Verne over forty years ago, but his idea stuck in my subconscious. To return to more ancient geometers, we know that Plato was very interested in geometry, and Vitruvius gives one of Plato's theorems, on the doubling of a square. For example, given a square with ten-foot sides, and therefore an area of one hundred square feet, how can one construct a square with an area of two hundred square feet? Our Figure 5

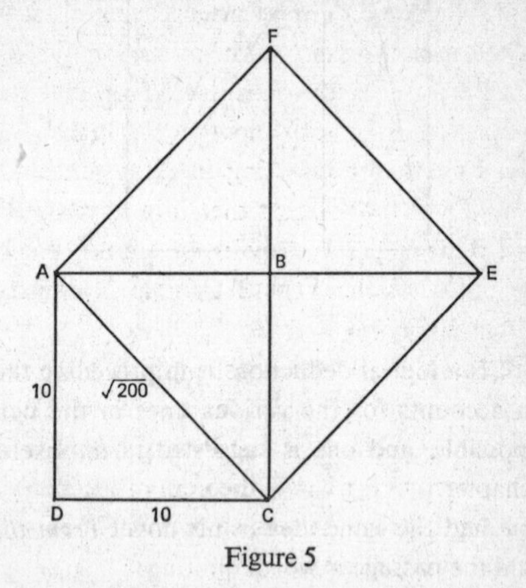

Figure 5

shows how this can be done, and the correctness of the construction is evident, since the new square consists of four triangles, each of which is *congruent* (identical) with a triangle which is half of the original square.

Vitruvius remarks that this problem cannot be solved by arithmetic, referring to the discovery, which caused a crisis in Greek mathematical development, that if the sides of a square are of unit length, the length of the diagonal cannot be expressed as a rational number, that is as a fraction p/q, where both p and q are integers. By Pythagoras' theorem, the square on the diagonal is equal to the sum of the squares on two sides of unit length, so that $(\text{diagonal})^2 = 2$, and the length of the diagonal is equal to $\sqrt{2}$ units of length. It can be shown, fairly simply, that $\sqrt{2}$ cannot be a rational number, and the Greeks knew no other kinds of number at that time. (See p. 155 for the proof.)

There is a converse to Pythagoras' theorem, which says that if we have a triangle with sides of length a, b and c, and $a^2 = b^2+c^2$, then the triangle is a right-angled triangle. Now the simplest such triangle has sides 5, 4 and 3 units in length, where by 'simplest' we mean a triangle whose sides are of integral length, and where the integers are the smallest possible. The ancient civilizations which preceded the Greek civilization must have been able to construct right angles, but we do not know how they did it, and one can be severely taken to task for asserting that the ancient Egyptians knew that $5^2 = 3^2+4^2$, and must therefore have used a knotted string to form right angles. Vitruvius, of course, knew this, and mentions the utility of the 5, 4, 3 triangle in constructing the treads of staircases (Figure 6).

Vitruvius remarks that Pythagoras is said to have sacrificed an ox to the Gods when he discovered the theorem named after him. On the other hand, Cicero says that Pythagoras invariably sacrificed one ox on discovering a new theorem in geometry, but when he discovered *the* theorem, he was so overwhelmed with joy that he sacrificed one hundred oxen!

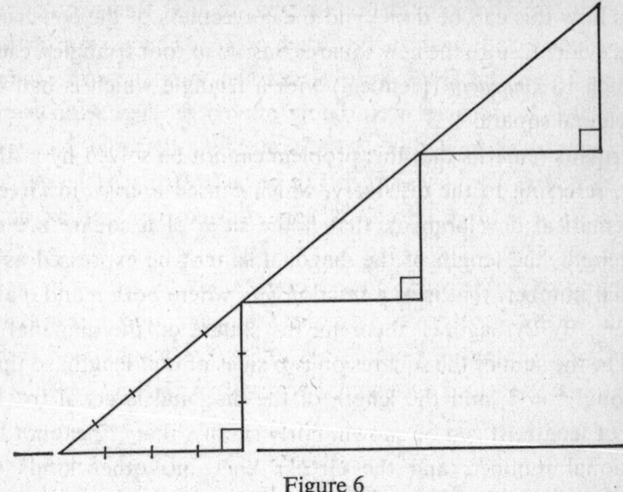

Figure 6

The tendency to celebrate, on making a fine discovery, is fairly universal among scientists, but most of the mathematicians I know believe that the kingdom of God is within themselves, and therefore do something which pleases the inner man when they complete a good piece of research. Those who like cigars buy themselves an exceptionally good cigar, and so on, bribing the inner consciousness, or the subconscious, whatever it may be, to produce more good results. And mathematicians were doing this a long time before Skinner evolved his theories of reinforcement.

Vitruvius brings in Plato's little theorem as an illustration of the permanent good which flows from the actions of such great men, as compared to athletes, who leave nothing behind them. It obviously rankles with Vitruvius that in his time athletes were richly rewarded, far more than philosophers. *Plus ça change . . .*

Plato solved the problem of the duplication of a square, but there is another geometrical problem of great renown called 'the duplication of the cube' which is not solved so easily. Vitruvius tells the story relating to this problem.

On the occasion of a severe plague, the Delphian oracle was

asked how the Gods could be placated, so that they would relinquish their wrath against the Greek people. The answer of the oracle was that the Gods were dissatisfied with the size of the altar on which sacrifices were being offered to them, and were demanding an altar of double the size. Since the altar was the shape of a cube, the new altar had to be built with edges equal to ka, where a was the length of an edge of the original altar, and $k^3 = 2$.

A young and rather obscure poet commemorated the event by a poem in which he described how the Athenians built a new altar in which the dimensions of the old altar were all doubled. The poet became notorious overnight, because everyone in Greece was laughing at his capital ignorance in matters of geometry, in not knowing that doubling the dimensions of the altar would give a new altar not twice, but *eight* times as large . . .

We have, of course, advanced since the days of the ancient Greeks, but some time ago, when I was having some ceramic figurines fired, I was told that the cost would be so much a square inch. When I inquired how the total number of square inches in the figurine would be calculated, I was told that the length would be multiplied by the breadth! The great Jonathan Swift, in his early eighteenth-century work, *The Marvellous Travels of Lemuel Gulliver*, makes a number of mathematical observations, and is never wrong. During the First Voyage, Gulliver falls among the *Lilliputians*, who were minute creatures, and the following extract is very relevant to our discussion:

The Reader may please to observe, that in the last Article for the Recovery of my Liberty, the Emperor stipulates to allow me a Quantity of Meat and Drink, sufficient for the support of 1728 *Lilliputians.* Some time after, asking a Friend at Court how they came to fix on that determinate Number; he told me, that his Majesty's Mathematicians, having taken the Height of my Body by the Help of a Quadrant, and finding it to exceed theirs in the Proportion of Twelve to One, they concluded from the Similarity of their Bodies, that mine must contain at least 1728 of theirs, and consequently would require as much Food as was necessary to support that Number of *Lilliputians*. [The reader will notice that 1728 = 12 . 12 . 12.]

The Greeks did solve the problem of the duplication of the cube, by inventing new mathematical curves (p. 246). Interestingly enough, when Plato was asked about the meaning of the Delphian oracular utterance, he said that the Greeks shamefully neglected the pursuit of geometry, and the oracle wanted to call attention to this fact. Where is the modern equivalent of the Delphian oracle?

Vitruvius, as an architect, was greatly concerned with the sites of buildings, their appearance, and the influence of the weather on his constructions. He also has some observations on the influence of climate on people.

Races bred in the north, he says, are of great height, have fair complexions, straight red hair, grey eyes and a great deal of blood. (He must have been talking about the Celtic races. The Romans occupied Britain for several hundreds of years.) The races nearest to the southern half of the earth's axis and that lie directly under the sun's course are of lower stature, with a swarthy complexion, hair curling, black eyes, strong legs, but little blood, which makes them over-timid to stand up against the sword, but great heat and fevers they can endure without timidity. Hence, men that are born in the north are rendered over-timid and weak by fever, but their wealth of blood enables them to stand up against the sword without timidity . . .

The truly perfect territory, situated under the middle of heaven, and having on each side the entire extent of the world and its countries, is that which is occupied by the Roman people. Exactly as the planet Jupiter is itself temperate, its course lying midway between Mars, which is very hot, and Saturn, which is very cold, so Italy, lying between the north and south, is a combination of what is found on each side, and her pre-eminence is indisputable. And so by her wisdom she breaks the courageous onsets of the barbarians . . .

The Roman Empire was, of course, of a vast extent at that time. Even if it did not cover the whole of Africa, there must have been some Dinka among the slaves brought to Rome, and they are very tall. Vitruvius states that all races nearest to the southern half of the earth's axis are of lower stature. Nobody stands up to great heat and fevers 'without timidity'. In many parts of Africa the sun is regarded as an enemy, and Africans tend to cast a jaundiced eye on the sunworship of Europeans visiting Africa. But Vitruvius obviously favours symmetry in statements as well as in design, and his statement is well-rounded, with an anticipated conclusion.

To return to the appearance of buildings, Vitruvius was well aware of the effects of refraction by the atmosphere. He believed, as people did for centuries after him, that rays of light streamed from the eye, rather than impinged on the eye from the object viewed, but it is clear all the same that he knew all that it was necessary for an architect to know about art and illusion. This is shown in his detailed comments on the design of temples. All temples were designed with columns, and these were tapered towards the top to give the illusion of greater height. In Figure 7 a simple geometrical instrument for producing this tapering effect, in a drawing, is shown in action. This ancient device will be discussed in our chapter on curves, p. 244.

Again, Vitruvius remarks that when we stand in front of a building, the upper part looks as if it were leaning back. This is true, although we may not accept his explanation that this is because a line drawn from the eye to the top is longer than one drawn from the eye to the bottom. To counter the effect produced by height, Vitruvius suggests that all the parts of a building which are to be above the capitals of the columns, the architraves, the friezes, cornices, tympana, gables and so on should be inclined to the front a twelfth part of their own height, and asserts that when this is done they will appear plumb and perpendicular to the beholder.

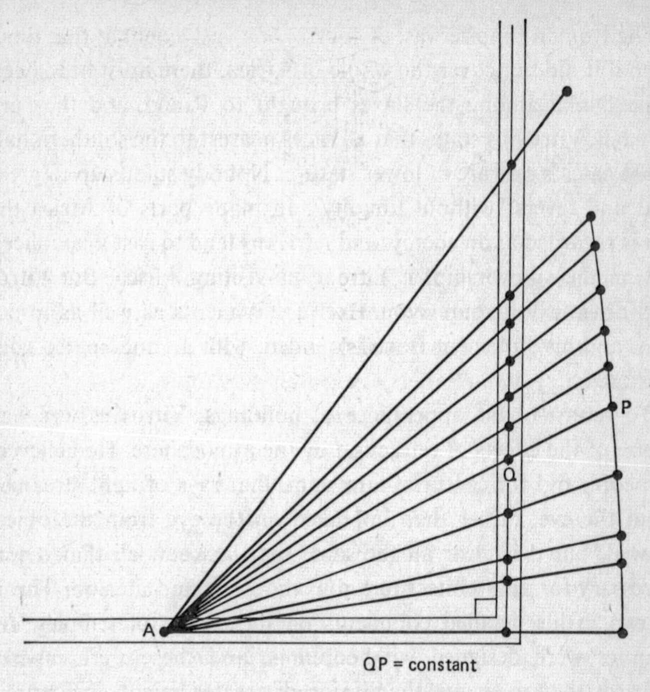

Figure 7

Vitruvius obviously copies the Greeks in his careful production of optical illusions in architecture, and indeed he is always very respectful when he talks about the ancients, but he does hint that Roman architecture in his time was rather good.

Columns, in Greek and Roman architecture, are of three kinds, Corinthian, Ionic and Doric, and Vitruvius gives all the necessary details of their history and design. They are easily recognized by their capitals (Figure 8). Before we embark on his description, let us consider a method for the approximate construction of spirals, which occur in the Ionic order. This method is given by Dürer, in his book on practical geometry (p. 44), but must have been known to Vitruvius. A spiral is constructed as a series of circular arcs (Figure 9). The centre P_1 gives a quadrant

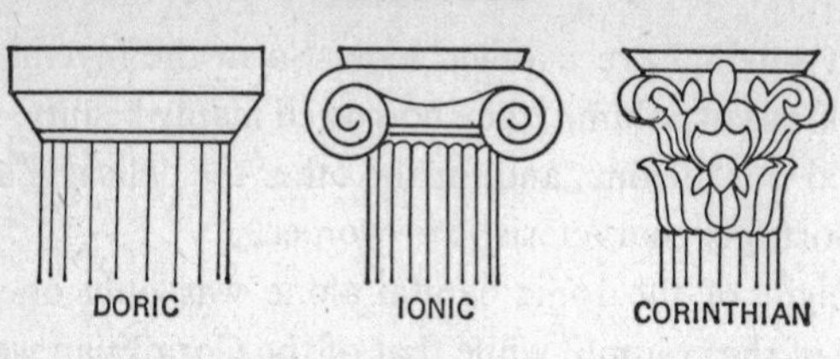

Figure 8

of a circular arc, P_2 is the centre of the continuing circular arc, P_3 is the next centre, and so on. Each arc is a quadrant, and the successive radii of the arcs diminish in any desired manner. We shall discuss spirals in more detail in our chapter on curves.

The *Doric*, says Vitruvius, was the first architectural order to arise. Legend has it that they measured a man's foot, found it was one sixth of his height, and applied this ratio to erecting a whole shaft, including the capital. Thus the Doric column, which exhibits the proportions, strength and beauty of the body of a man . . . Then afterwards, when they desired to construct a temple to Diana, in a new style of beauty, they translated these footprints into terms characteristic of the slenderness of women, and made a column the thickness of which was only one eighth of its height,

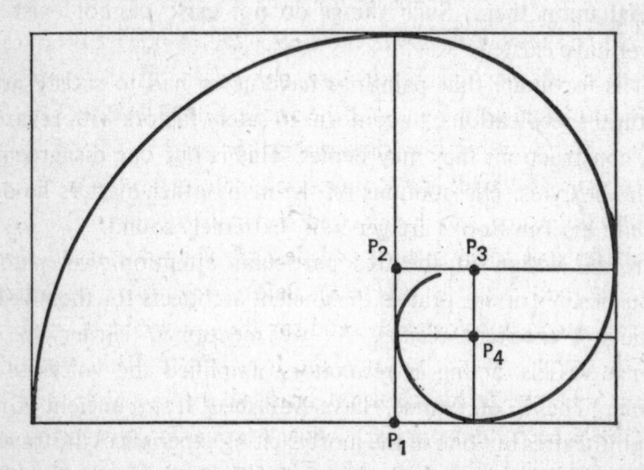

Figure 9

so that it might have a taller look. So in the invention of two different kinds of column, they borrowed manly beauty, naked and unadorned, for the one, and for the other the delicacy, adornment and proportions characteristic of women.

The height of the Ionic capital alone was only one third the thickness of the column, while that of the Corinthian was equal to the entire thickness. The Corinthian column was evidently based on the acanthus, and Vitruvius tells the story of Callimachus passing the tomb of a young girl of Corinth, and seeing an acanthus growing, with the stalks pressed out into spiral scrolls by a roof-tile which covered the girl's possessions. He was so delighted with the pattern that he incorporated it into some columns he was designing (Plate 6).

The acanthus is probably the most overworked design in the whole history of decorative art. When Vitruvius discusses fresco painting, he complains bitterly that the painters of his time were decadent, painting monstrosities rather than the truthful representations of definite things. 'For instance, reeds are delineated in the place of columns, with numerous tender stalks and volutes growing up from the roots and having human figures senselessly seated upon them. Such things do not exist, cannot exist and never have existed!'

It is fortunate that paintings have never had to satisfy architectural specifications, to conform to safety-factors with regard to any constructions they may depict. This is our one disagreement with Vitruvius. His opinions on the many other matters he deals with in his Ten Books are generally extremely sound.

In the design of theatres particular attention was paid to acoustics. Vitruvius praises the ancient architects for their skill in achieving excellent results. As we mentioned earlier (p. 17), bronze vessels, acting as resonators, amplified the voices of the actors. These, of course, have vanished from ancient Greek amphitheatres but one of the most exciting experiences the traveller can have in Greece nowadays is to test the perfect acoustics of the amphitheatre at Epidaurus. Standing on the rim of this vast bowl, one can hear a whisper at the centre of the stage. This, of course, may well be accidental good design, but certainly Vitruvius was deeply concerned that an audience should be able to hear, and was ready to use any means known at that time to achieve this purpose.

Some modern theatres, with a round stage almost certainly based on a misunderstanding of what the Shakespearean stage was like, have no amplifying devices, with the result that it is only possible to hear about one quarter of what the actors are saying, since the producer dutifully directs his cast to address four corners of the theatre, in sequence.

We now give Vitruvius' design for a Roman theatre (Plate 4). Having fixed on the principal centre, draw a circle to represent the inner perimeter of the rows of seats, and in this circle inscribe four equilateral triangles at equal distances apart, as the astrologers do, says Vitruvius, when constructing a figure of the twelve signs of the zodiac. (The twelve points are simply spaced round the circle at angular distances of thirty degrees from each other.) The outer circle denotes the outer perimeter of the theatre, where there was a colonnade, for walking during the intermissions. The orchestra then was simply a semicircular area in front of the stage, where the chorus and possible instrumentalists could operate, and it also contained seats for privileged people. In the Vitruvius design the orchestra is the upper half of the inner circle, and *DD* is the line which separates the orchestra from the proscenium *EEE*. The radii from the centre of the circle to the vertices of the upper triangles give the directions of entrances to the orchestra, and of staircases leading up to the rows of seats. The line *FF* is the face of the scene itself, which we call the stage nowadays, and *GG* represents the back of the stage, the postscenium. The machines for effecting a change of scene are at the points marked *O*.

There is a reference in Vitruvius to triangular pieces of machinery which can revolve, each piece having three decorated faces.

When a play was to be changed, or when gods entered to the accompaniment of sudden claps of thunder, these machines revolved, presenting a different face to the audience.

There were tragic, comic and satyric scenes. Tragic scenes had columns, pediments, statues and other objects suited to kings. Comic scenes exhibited private dwellings, with balconies and rows of windows, and satyric scenes were decorated with trees, caverns, mountains and other rustic objects, in landscape style. Tragic scenes apparently never involved ordinary people. Satyric scenes involved satyrs, woodland deities who attended Bacchus, and were usually represented as having pointed ears, short horns, the head and body of a man and the legs of a goat. They were fond of riotous merriment, and goat-like in their behaviour, and there is recently discovered evidence that these satyric scenes involved the throwing of a lot of horse manure, whether simulated or real we do not know.

Vitruvius points out that in the design of a Roman theatre the platform of the stage had to be made deeper than that of the Greek theatre, since all Roman artists performed on the stage, while the orchestra contained the places reserved for senators.

The Greek theatre was designed on the basis of three squares and three centres (Plate 5), and the orchestra was roomier, the front of the stage was set farther back, but the stage itself had less depth. The tragic and comic actors performed on the stage, while other artists performed in the orchestra itself.

At the beginning of this chapter we mentioned that Vitruvius thought that an architect should be acquainted with astronomy and the theory of the heavens. By 'theory of the heavens' it is clear that Vitruvius means *astrology*, which he says was derived from the Chaldeans, and it was clear to him that the forces which are exerted by the various dispositions of the heavens must be taken into account by a good architect. He also discusses the construction of sundials, and in this connection he mentions the fifteenth part of the circumference of a circle. This subtends an angle of 360/15 = 24 degrees at the centre of the circle. We still use a Babylonian number system for measuring degrees. If the measurement of angles is ever changed into the metric system, which unit of angle will be called 100? Will it be the angle on a line, at present 180 degrees, or will it be the angle of a complete rotation, at present 360 degrees? In view of what happened to the British pound, when Britain changed to decimal currency, we cannot be certain. The irrational often wins.

How did Vitruvius obtain an angle of 24 degrees, using straight edge and compasses? He knew his Euclid, and would be able to construct a regular pentagon, which has five sides, in a circle. Each side of this figure subtends 360/5 = 72 degrees at the centre. The construction of an equilateral triangle gives an angle of 60 degrees, and subtracting half of 72, which is 36, from 60 produces 24. We shall have more to say about Euclid in Chapter 6.

We now come to some of the machines described by Vitruvius, which are depicted very lovingly in the beautiful French edition we have been using. Archimedes had already described an odometer, a device for measuring the distance gone by a chariot, and Vitruvius also describes one. He assumes that the wheels of the chariot are four feet in diameter, so that if a wheel has a mark made upon it, where it touches the ground, and the chariot moves until the mark first comes down to the ground again, then, says Vitruvius, the chariot will have moved twelve and a half feet. This gives a value for the fundamental constant π, the ratio of the circumference of a circle to its diameter, equal to 3·125, which is rather low. Archimedes, whose work Vitruvius discusses with approval, certainly used far better approximations than this, and there is a possibility that the Vitruvius text used for the French translation may be faulty at this point.

The odometer described by Vitruvius was a device attached to the axle of the chariot such that when a wheel made one revolution, a drum with four hundred teeth on it moved round one notch. A complete rotation of the drum was equivalent to a motion of 400(12·5) = 5,000 feet by the chariot. At this point the mechanism dropped a stone into a bronze bowl, and if 5,000 feet was one Roman mile, the number of stones in the bowl at the end of the journey gave the distance traversed in Roman miles.

Except for the bronze bowl and the stones, a similar principle is used today. Another device which used gears, and which is no longer in general use is a water clock, and here the 1684 edition of Vitruvius gives a fine picture (Plate 8), with a full description, which is well worth repeating.

Water flows steadily up the pipe marked *A*, and issues as *tears* from the eyes of the cherub on the right, who is attached to the pipe. The water, as the tears accumulate, flows through the hole marked *M* near to *A*, and into the pipe *CD*. This contains a float, and as the water level rises in *CD*, the float rises and pushes up the cherub with the wand, on the left. It takes 24 hours for the cherub to move up the column with the hours marked on it.

But note that water also enters the pipe *FB*, and when the float is at the top of the pipe *DC*, water has also reached the top of the pipe *FB*, and begins to flow down the pipe *BE*, producing *a siphon action*. This siphon action sucks all the water out of the pipe *FB*, and this water flows down onto a mill-wheel marked *K*, turning it one sixth of a revolution. In the meantime, the cherub on the left holding the wand has slid down from his elevated position. The gears down below are so devised that the column turns through 1/366 of a revolution when the water pours down every 24 hours.

Besides this difference between our concept of the number of days in a year and what a reasonable number was thought to be in France in 1684, the division of the hours on the rotating column of the water clock are not uniform, as nowadays, but designed to distinguish between the hours of day and night at various seasons of the year.*

* Vitruvius himself, when discussing water clocks, mentions the division of a year into 365 days. This came in with the Julian calendar in 46 B.C. Perhaps the difficulty of adjusting to leap years accounts for the French clock.

We must not leave Vitruvius without mentioning the fact that he describes various kinds of water-lifting devices, which are still used in some countries, including the one called the Archimedean screw (Plate 9). This was still used in the Sudan, when I was there in the 1950s, and is obviously one of the simplest fool-proof devices for raising water from one level to a higher one.

Finally, Vitruvius describes many war-machines, and perhaps the most striking illustration in the 1684 edition of Vitruvius is that of a siege tower (Plate 7), twenty stories high, with a staircase running up the middle.

Roman victories, such as Julius Caesar's campaigns, as anyone who reads a description of them soon remarks, depended to a large extent on engineering feats which the enemy forces thought impossible, throwing bridges across rivers thought to be unfordable, and so on. Reading the bluff, honest Vitruvius, one senses the existence of a remarkable technology, with a very sound perception of practical geometry.

Exercises

1. Using compasses and ruler, erect an equilateral triangle ABC on a given line AB. Erect another, ABD, and observe that CD bisects AB.

2. Draw a circle, centre O, and let A and B be two distinct points on the circumference. Erect an equilateral triangle ABC on AB. Join OC. What can you say about the line OC in relation to the arc AB; in relation to the angle AOB?

3. Draw a circle, centre O, and choose any point A on the circumference. With centre A and radius AO draw a circle, cutting the original circle in B and F. With centre B and radius BO draw another circle, cutting the original circle in C and A. Continue in this manner, until you obtain a regular hexagon $ABCDEF$ inscribed in the circle, and simultaneously a design for a rose window.

4. The previous exercise has shown that you can obtain a regular hexagon $ABCDEF$ by *stepping round* the circle, beginning with any point A on the circumference, each step being equal to the radius of the circle. You also see that in this way you can move from one end of a

diameter A of the circle to the other end D without the help of a ruler. Draw such a regular inscribed hexagon, and bisect the arcs AB, BC and CD, joining the midpoints to O, and producing the lines obtained until the circle is reached again. You now have the vertices of a regular inscribed duodecagon (12-gon). Repeat the construction of the previous exercise, that is, taking the vertices of the 12-gon in turn, draw circles with these vertices as centres with radii equal to that of the original circle. You will obtain another design for a rose window.

5. Draw a circle, centre O, and let AE be a diameter. Bisect the arc AE at C, and bisect the arcs AC and CE at B and D respectively. Produce the lines OB, OC and OD to meet the circle again. You now have a regular inscribed octagon (8-gon). With centre A and radius AB draw a circle. Continue in this way all round the vertices of the octagon.

6. Draw a circle, centre O, and inscribe a regular pentagon $ABCDE$ in it. (A construction with ruler and compasses will appear in Chapter 2 (p. 68), but in the meantime use the fact that each side, such as AB, subtends an angle of 72° at O.) With centre A and radius AB, draw a circle. Continue thus around the vertices of the pentagon. This produces another window design.

7. Three points O, A and B are given, where $AO > AB$, and we wish to draw a circle, centre O and radius AB, without transferring a length with our compasses. Construct an equilateral triangle ADO on the side AO. Let the circle centre A and radius AB cut AD in F, where F lies between A and D. Let the circle centre D and radius DF cut the line DO in G, where G lies between D and O. Prove that the length OG is equal to the length AB, so that the circle centre O and radius OG satisfies our requirements. Compare this construction with that given on p. 22. Examine the case $AO < AB$.

8. Draw a regular pentagon (see Exercise 6), and sketch a man (or woman) inscribed in the pentagon, that is with head, hands and feet at the five extremities.

9. Draw two spirals by Dürer's approximate method (p. 34), using compasses, and incorporate your spirals in a drawing of an Ionic column. (It is sufficient to go as far as P_4, and then to complete the volutes freehand.)

10. $ABCD$ is a square of side $a+b$, and P is a point on AB distant a

from A, and therefore b from B, Q is a point on BC distant a from B and b from C, R is a point on CD distant a from C and b from D, and finally S is a point on DA distant a from D and b from A. Show that $PQRS$ is a square, by verification, if you do not know how to prove it. Using the theorem that the area of a right-angled triangle of sides a and b containing the right angle is equal to $(ab)/2$, and that the area of a square of side $a+b$ is equal to $(a+b)^2$, find the area of the square $PQRS$. Deduce the theorem of Pythagoras that the square on the hypotenuse of a right-angled triangle is equal to the sum of the squares on the other two sides. (You will have to use the algebraic identity: $(a+b)^2 = a^2+2ab+b^2$.)

·2·

ALBRECHT DÜRER

ALBRECHT DÜRER lived from 1471 to 1528, and it may seem odd that we should leap from Vitruvius to Dürer in our discussion of geometry and the liberal arts, but there are sound reasons for doing this. Dürer has often been chosen as the German incarnation of the Renaissance, which coincided with the Reformation in Germany. Dürer was not only a fine geometer, he was also a thinker on many matters outside geometry, and his art was of the greatest importance to him. In fact, he never lost sight of those fundamental problems which, later on, were to become the domain of what we call Aesthetics, the theory of Beauty.

Another important reason for choosing Dürer is that we know a lot about him. He wrote two treatises, one on descriptive geometry called *Underweysung der Messung mit dem Zirckel und Richtscheyt*, a treatise of constructions with the compasses and ruler, and the other: *Vier Buecher von Menschlicher Proportion*, four books on human proportions, both part of a far larger programme, which he did not live to finish. We shall refer to the first book as the *Underweysung*, and to the second as the *Vier Buecher*.

We also have letters he wrote to his friend Wilibald Pirkheimer, which cover the periods when Dürer had left his native Nuremberg to visit Venice, and then Bologna, to learn more of the art of perspective, and a fairly detailed diary he kept when he took his wife and her maid on a visit to the Netherlands.

Pirkheimer was rich, whereas Dürer had grown up as the son of a goldsmith, an artisan, and was often worried about money.

So part of the correspondence with Pirkheimer deals with money problems, but part also with Pirkheimer's affairs of the heart. The letters are delightful, and give one a warm feeling towards both Dürer and Pirkheimer. Pirkheimer was also a scholar, whereas Dürer had very little Latin, and Dürer evidently made good use of Pirkheimer's library, which still exists. Pirkheimer also helped Dürer by translating works which Dürer would not have been able to read by himself. The relationship is one of the great ones in the history of art, comparable to, but not as tragic as that between Vincent van Gogh and his brother Theo. It was Pirkheimer who saw the *Vier Buecher* through the press after Dürer's death, and a flavour of Dürer's character is given by the letter written to Pirkheimer some time earlier. The four books were to be dedicated to Pirkheimer, and Dürer wanted him to write a preface. Dürer wrote:

Dear Master,

Will you have the goodness to arrange the preface as I here indicate?

1. I desire that nothing boastful or arrogant appear in it,
2. That there be no reference to Envy,
3. That nothing be mentioned which is not in the book,
4. That nothing be introduced which is stolen from other books,
5. That I write only for our German youth,
6. That I give the Italians very high praise for their naked figures, and especially for their Perspective,
7. That I pray all such as know anything instructive for art to publish it.

Despite these specific instructions, Pirkheimer enlarged in his preface upon several matters which Dürer had asked him not to mention. It seems that it was a standard custom to warn critics not to let jealousy of the work they were reading carry them away, and Pirkheimer warns them. He also spoke of other matters which disturbed Dürer, so Dürer wrote to Pirkheimer again, and the following extract from the letter gives a good idea of Dürer's intentions in writing his books:

Now, as this book deals with nothing but Proportion, I desired to keep all references to Painting for the book which I intend to write upon that subject. For this Doctrine of Proportions, if rightly understood, will not be of use to painters alone, but also to sculptors in wood and stone, goldsmiths, metal founders, and potters who fashion things out of clay, as well as to all those who desire to make figures.

Dürer rewrote his dedication to Pirkheimer, and introduced into it the matter which he had asked Pirkheimer to put into the Preface, but Dürer did not live to see the book published.

Dürer's books were often reprinted, and he was widely known as both a writer and an artist. It was fortunate that he had not had the traditional education of an upper-class German, and therefore little Latin, so that his German was not based on the convoluted Latin of Cicero, but derived directly, like that of Martin Luther, from the German of the man in the street. When Dürer had to invent words for mathematical concepts, he used the words which had always been used by the people he knew, the artisans of Nuremberg, to express similar notions. As a result he is delightful to read, the language being simple and powerful, and, of course, comprehensible.

We have already mentioned Dürer's visit to Bologna to learn more about perspective, the art of representing space on a flat surface. During the Middle Ages this problem had been ignored, as pictures of this period in any art gallery indicate. Such pictures are spread out flat on the canvas, or so it seems to us, and the effect is often very striking and has been reproduced by some modern painters. Some Matisse interiors are painted in this style.

During the fourteenth century, however, the idea occurred to many artists that a canvas can be thought of as a window through which the observer looks out onto a part of the world. Matters of observation were then easily depicted. Objects diminish in size as they move away from the spectator, and the parallel sides of a floor or a ceiling recede towards the end of a room as you look at them. In particular lines at right angles to the canvas (called *orthogonals*) which lie in a plane, such as the sides of a furrow in a ploughed field, look as if they all pass through the same distant point on the horizon.

About 1420 there was a great advance when the sides of a cube which were not orthogonals were depicted as converging towards *two* points located on the horizon, also called *the vanishing line* in the theory of perspective (Figure 10). In the 1450s it was also dis-

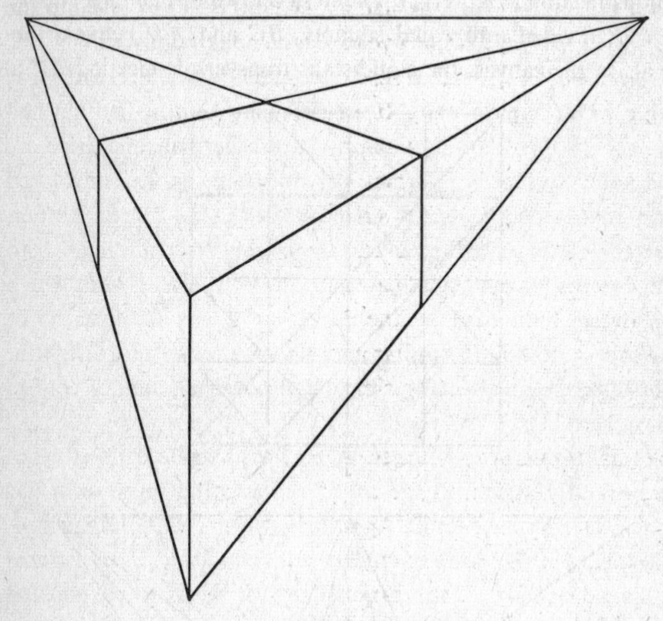

Figure 10

covered that *all* orthogonals, not only those in the same plane, had to be drawn converging to the same *vanishing point*, and that this one point was sufficient to establish the vanishing line, which was the horizontal line on the canvas through this point.

Earlier, the problem of depicting *equidistant transversals* occupied artists. You have your canvas set up in a room with a tiled floor. One set of edges of the square tiles are orthogonal to your canvas. The other set form parallel lines across the floor, *transversals*, which are equidistant from each other. How do you represent these *equidistant transversals* on your canvas? An incorrect rule-of-thumb method was still in vogue in 1435 by which the intervals on the canvas between one transversal and the next were to be diminished by one third.

By the fifties this problem had been solved in a correct and ingenious fashion (Figure 11). The floor shown in plan is a square, $ABCD$, divided into equal squares, BC and AD being orthogonals to the canvas, the equidistant transversals meeting BC at

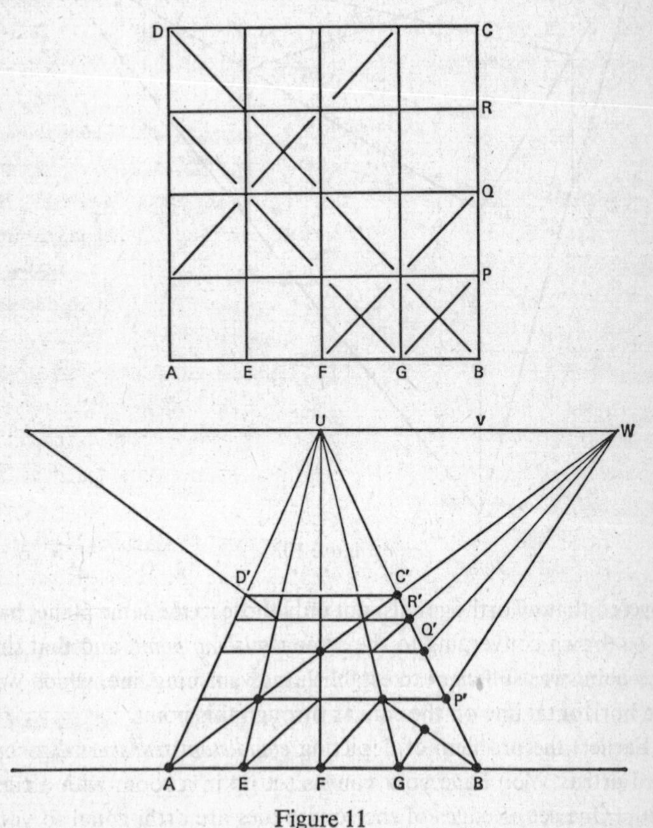

Figure 11

P, *Q* and *R*. In the picture, the orthogonals will all meet at a point, which we call *U*, and the transversals will remain parallel to *AB*, which we may assume is unchanged. We notice that each of the sixteen small squares into which *ABCD* is divided has two diagonals, and one set of diagonals of these small squares are parallel to each other, and so are the other set.

If we draw a line through *U* parallel to *AB*, this line *v* is the *vanishing line*, and the set of diagonals parallel to *FQ* are represented on the canvas by a set of lines which pass through a point *W* on *v*. This is sufficient to enable us to draw the image of the transversals, although it is also true that the other set of parallel diagonals are depicted by a set of lines passing through another point on *v*.

To find the transversal corresponding to the one through *P*, we note that the points *A*, *E*, *F*, *G*, and *B* are unchanged in the picture. Join *GW*, and let *P′* be the point where this line meets *BU*. Then the line through *P′* parallel to *AB* depicts the transversal through *P*.

To find the transversal corresponding to the one through Q, we join *F* to *W*, and where this line meets *BU* gives us *Q′*, and the parallel to *AB* through *Q′* is the representation of the transversal through *Q*. If the drawing be made, and if *U* is on the perpendicular bisector of *AB*, it can be verified that while one set of diagonals in the drawing pass through *W*, the other set pass through the point on *v* equidistant from *U*.

Dürer's work after about 1500 shows an excellent grasp of perspective, but he was always anxious to improve his art, and especially to learn something of the theoretical foundations of perspective. No book on the subject had appeared as yet, but the theory, usually attributed to the great architect Filippo Brunelleschi, had been known to a small circle in Italy for over fifty years. Dürer therefore left Venice, where he was staying in 1506, to travel to Bologna, over a hundred miles away, in order that he might learn this 'secret' art.

Part of our purpose in this chapter is to show that the theory of perspective arises easily from certain geometrical considerations. But practical perspective drawing arose naturally, just as it is taught nowadays, as a set of rules. It had nothing to do with the mathematical analysis of the process of vision known as *Optics* in antiquity, and as *Perspectiva* in the Latin Middle Ages. As we shall see in Chapter 5, when we discuss Euclid's *Optics*, this discipline was discussed by Euclid to a rather limited extent, when he attempted to express in geometrical theorems the relation which exists between the so-called reality of objects and the impressions which make up our visual images.

Euclid postulated that objects are perceived by straight rays which converge in the eye, so that the visual system can be thought of as a *pyramid*, with the eye as its vertex and the object as its base. Euclid also introduced another postulate, that the apparent size of any object depends on *the angle* subtended by the object at the eye. There is much more, of course, to the theory of vision, and some of it is still controversial.

But to return to Euclid, his theory of optics made no attempt, as far as we know, to deal with the problems of artistic representation. For centuries the idea of *intersecting* the pyramid of sight with a plane, the canvas in fact, did not emerge. Obvious as this idea may seem to us, with projectors, lamps shining through the darkness and so on, it was left to Brunelleschi to formulate the great notion that a picture on a canvas is obtained by taking a section, with the canvas, of the pyramid or cone which joins the one-eyed artist to his subject. Perspective is the theory of the one-eyed artist's vision.

It is curious that this idea does not occur anywhere in Euclid, but we know that many of his works have disappeared without trace. We shall give a fairly complete discussion of the theory of perspective later on (p. 52). Dürer was a fine enough geometer to have been able to understand anything the Italians were able to teach him, and it seems that his visit to Bologna was a satisfactory one. Problems of perspective arise in the works of Dürer which date from between the years 1510 and 1515, the period of one of his most famous works, *Melencholia I* (Plate 10).

Melancholy, originally considered to be the least desirable of the four humours, caused by blood, phlegm, yellow bile or black bile, became identified during the Renaissance with the artistic temperament, or *creativity*, to use a much overworked modern term. Dürer's representation is surrounded by all the attributes of geometry and the constructive arts, yet she seems to be reduced to impotence and despair by her awareness of realms of thought inaccessible to her. A modern iconographer could discuss this one engraving at great length, since every object in the picture is also a symbol, but we pass on to the equally famous *St Jerome in his Cell*, which is of the same period (Plate 11).

Dürer appears to have felt that this engraving went with the *Melencholia I*, for he often presented them or sold them together. They can be considered, perhaps, to represent *sacred and profane learning*, although as with pictures entitled *sacred and profane love*, one might argue about which symbol represents the sacred, and which the profane. In any case, at a recent sale at the great print shop, Colnaghi's, in Bond Street, London, the *Melencholia I* was offered for £4,500 and the *St Jerome* separately for £8,500. This would have certainly dismayed Dürer, who was often worried about money. At one period of his life the Emperor Maximilian, with whom he was on very good terms, asked the Town Council of Nuremberg to allow Dürer to live tax-free, but the Council was reluctant to make exceptions, and Dürer wisely abandoned the idea. Soon after Maximilian granted Dürer a pension, the Emperor died, and his successor, Charles the Fifth, took some time to acknowledge his predecessor's obligations. Towards the end of his life Dürer asked the Nuremberg Town Council whether they would pay him what amounted to five per cent on a sum he deposited with them, and this they did, reducing it to four per cent for his widow when Dürer died.

Dürer deals with perspective as the culmination of his *Underweysung*, and gives some devices (Plates 12–14) for ensuring accurate perspective by mechanical rather than mathematical means. Some of these devices were already known, certainly to Leonardo da Vinci. The eye of the observer is fixed by a sight, and between it and the object there is inserted either a glass plate, or a frame divided into small squares by a net of black thread. In the first case, an approximately correct picture can be obtained by simply copying the contours of the model as they appear on the glass plate, and then transferring them to the panel or paper by tracing. In the second case the image perceived by the artist is divided into small units, whose content can easily be drawn upon a paper divided into a corresponding set of squares. Since it is undesirable that the artist's eye be never further from the glass plate than the length of his arm, the eye may be replaced by that of a large needle, driven into the wall, to which is fastened a piece of string with a pointer at the other end (Plate 14). The operator can then aim at the characteristic points of the object, and mark them on the glass plate, with the perspective situation determined by the needle, not by his moving eye.

We now consider perspective from the purely geometric point of view. Although the notion of the visual system as a cone, or pyramid, with the eye as the vertex, can be found in Euclid's *Optics*, it was left to Brunelleschi, nearly two thousand years later, to conceive a picture as being obtained *by intersecting the Euclidean pyramid with a plane*. The account given here is one which would come naturally to a geometer of the present century.

The point V represents the eye of the artist (Figure 12), the plane α' is a horizontal plane which we wish to depict in a picture, the plane α of which is taken to be perpendicular to α'. We are looking at points P' in the plane α', and depicting them by points P in α, and the line $P'P$ when continued must pass through the point V. We can say that we are *projecting* the points of α' onto the points of α, or vice versa, from the centre of projection V.

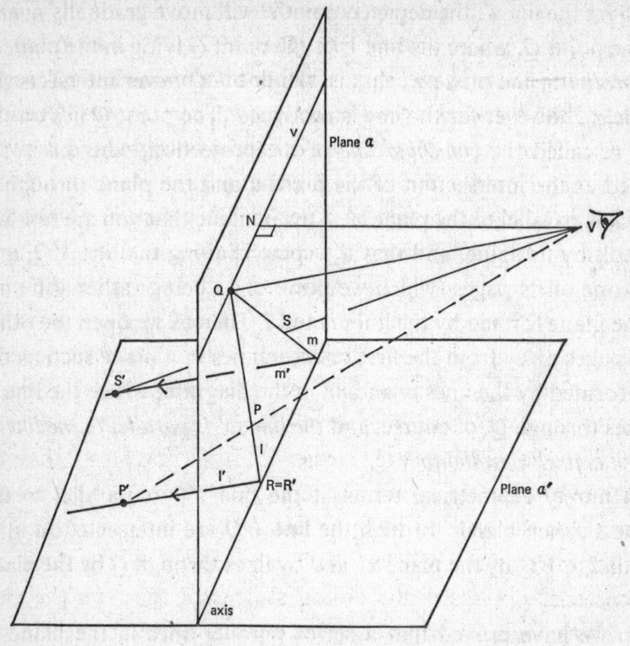

Figure 12

We call the line of intersection of the planes α' and α the *axis* of the projection, and we note that all points on the axis project into the same points, so that $R' = R$. This is because R lies in both planes α' and α. Now draw a line l' in α' which passes through R. If we join all the points P' on l' to V, and intersect these lines with the plane α, we obtain another line l in α which also passes through R. We can simply say that the join of the point V to the points of the line l' produces a plane, and the intersection of this plane with the plane α is a line. So the artist with one eye at V depicts a line l' lying in the plane α' by a line l lying in the plane of the canvas, α, and the original line and the depiction of it meet at a point on the axis.

Now, if we imagine P' to move away from the point R, always lying on the line l', the depicted point P will move gradually nearer to the point Q, where the line VQ, the point Q lying in the plane α, is *parallel* to the plane α', that is, the line VQ never intersects the plane α', however far the line is produced. The point Q lies on the line v, called the *vanishing line* of the projection, where v is obtained as the intersection of the plane α and the plane through V which is parallel to the plane α'. Now imagine that you are holding a book by its spine, and that the spine is along the line VQ, and that one of its pages (which we think of as being rather stiff) lies in the plane formed by the lines l' and l. Then as we open the other stiff pages away from the first one, each lies in a plane such as the one formed by the lines m and m' in the diagram, where the line m passes through Q, of course, and *the line* m$'$ *is parallel to the line* l$'$, *which is itself parallel to* VQ.

In purely geometrical terms, if the line VQ is parallel to the plane α', then planes through the line VQ are intersected in lines parallel to VQ by the plane α', and by lines through Q by the plane α.

So we have proved that a set of parallel lines in the plane α' must be depicted as a set of *intersecting* lines in the plane α, the point of intersection lying on v, the vanishing line, and that *to distinct directions in the plane α' there correspond distinct points on* v.

This simple statement corresponds to the evolution of the practice of perspective over a long period.

Suppose that we now drop a perpendicular VN onto the plane α, and that VQ makes an angle of 45° with VN. Then the angle NQV is also equal to 45°, since the angles of a triangle sum to two right angles, and therefore $NQ = NV$. Hence, if we find the point Q on v which corresponds to lines in the plane α' which make an angle of 45° with the axis, *the distance* NQ *is equal to the distance of the eye* V *from the canvas.*

If we look back at Figure 11, the distance UW represents the distance of the eye V from the canvas, the point U being the point

N in Figure 12, and the diagonals of the square making an angle of 45° with the axis AB.

Here we must point out that if we leave the horizontal plane α', and consider any point in space on the line VP', this point is also depicted by the one point P on our canvas. The line QP', which is a line in space, is depicted by the line QR on the canvas, and in fact any line in the plane defined by the three points Q, R and P' is depicted by the line QR. As soon as we try to represent objects in three-dimensional space on a two-dimensional plane, we lose the one-to-one property we have had so far, not counting points on the vanishing line v, which do not correspond to points in the plane α'.

But we can still say that *any line in space parallel to the line* VQ *in Figure 12 is represented by a line in* α *which passes through* Q. We know that in space of three dimensions any two parallel lines lie in a plane. Hence, if n is a line in space parallel to VQ, take a plane through VQ, and rotate it about the line VQ until it contains the line n. We now know that the join of V to points on n will give a line through Q in the plane α. This is the justification of the perspective drawing of a cube-like building in Figure 10. *There is just one point on the vanishing line* v *which corresponds to a set of lines in space which are all parallel to* α' *and to each other*, in the sense that all such lines are painted as passing through the same point of v.

Returning to Dürer, it can be said that he added nothing to the science of perspective as it was developed by the Italians. But the last section of the *Underweysung* is memorable in two respects. First, it is the first literary document in which a strictly representational problem received a strictly scientific treatment at the hands of a Northerner. None of Dürer's forerunners, and few of his contemporaries had any understanding of the fact that the rules for the construction of a perspective picture are based on the Euclidean concept of the visual pyramid, and it is a remarkable fact that even as late as 1546 treatises were reprinted which were partly empirical and partly completely wrong. In the second place the *Underweysung* emphasizes, by the position Dürer accorded his discussion of perspective, at the very end, that it is not a technical discipline destined to remain subsidiary to painting or to architecture, but an important branch of mathematics, capable of development. In fact, it did develop into projective geometry.

It must always be remembered that perspective is the science of seeing with one eye. When we begin to wonder what it is we actually see, a vast field opens before us, still assiduously cultivated by physiologists, physicists and philosophers. Let us return to one-eyed (*uniocular* is a nicer word) perspective.

Suppose that the large square $ABCD$ on our chequer-board floor in Figure 13 has a circle inscribed in it, which touches each of the four sides. When this circle is depicted on our canvas, we do not expect it to look like a circle. In fact it becomes an oval, the curve we call an ellipse. This ellipse will also touch the sides of the $ABC'D'$ figure. We show how to construct twelve points on this ellipse, enough to give a fairly good idea of its appearance. In Chapter 7, p. 186, we shall show how to make a perspective drawing of *any* circle.

The point O, the centre of the circle in Figure 13 becomes the point O', the intersection of the diagonals AC' and BD', and this is *not* the centre of the ellipse. The line AC' passes through the vanishing point W on the vanishing line v, and the line BD' passes through the vanishing point Z on v. All lines parallel to AC in the original figure become lines through W in the drawing, and all lines parallel to BD in the original figure become lines through Z in the drawing.

The ellipse will touch AB at E, and $C'D'$ at the point where UE meets $C'D'$. If we draw a line through O' parallel to AB, to intersect BC' at F' and AD' at H', the ellipse touches BC' at F' and AD' at H'.

To find other points on the ellipse, consider the point P where the line IK cuts the circle. Draw through P a line parallel to the

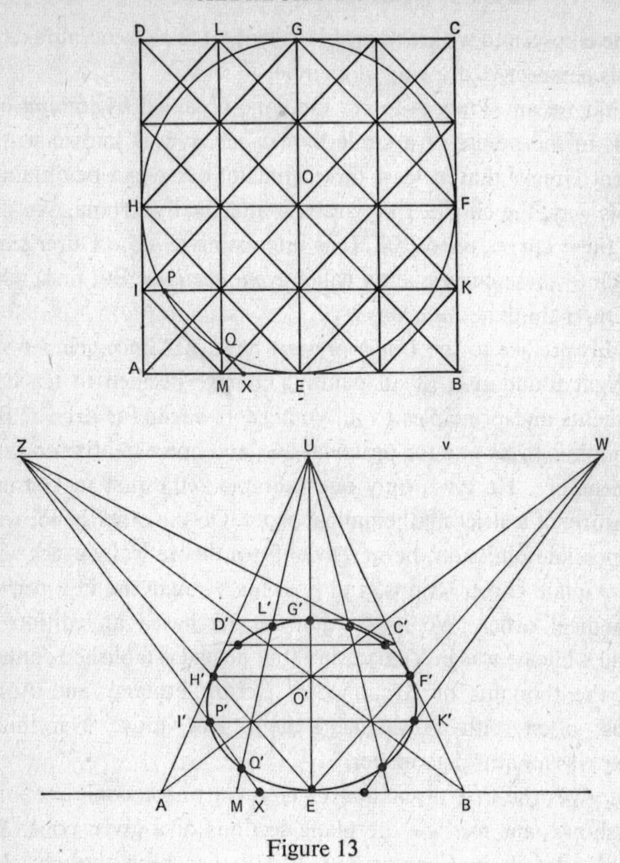

Figure 13

diameter BD of the circle, meeting AB in the point X. Then the point Q where this line meets LM also lies on the circle, since $OP = OQ$. To find P' and Q' in the perspective drawing, we note that the point X is unchanged, being a point on AB, so we can transfer the distance AX to the drawing and locate X. Now join X to Z, since PQ is parallel to BD, and P' is the point where ZX intersects $I'K'$, and Q' is the point where ZX intersects $UM' = UM$. In a similar manner we can find three other pairs of points on the ellipse, and we are thus able to make a fairly accurate sketch of this perspective drawing of a circle.

What we are doing is to *cut* the cone obtained by joining our eye *V* to the points of a circle by a *plane*. It was known to the ancient Greeks that at least three kinds of curve can be obtained in this way, the ellipse, the parabola and the hyperbola. We discuss these curves on p. 200. It is interesting to note Dürer's approach to these curves, aptly called *conic sections*. But first, what did Dürer think about *geometry*?

In his preface to the *Underweysung* he says: 'Since geometry is the right foundation of all painting, I have decided to teach its rudiments and principles to all youngsters eager for art . . . But the book is to be one for practical use, and not a treatise on pure mathematics. He gives only one example – the first in German literature, of a strict mathematical proof. On the other hand, with one possible omission, he never confuses theoretically exact with approximate constructions, and presents his material in a perfect methodical order. We know that he purchased an edition of Euclid while he was in Venice, and that he had established contact with the thoughts of Archimedes, Heron, Ptolemy and Apollonius, often with Pirkheimer's help. But above everything, Dürer was a natural geometer.

He gives the first discussion in German of the conic sections, and shows *how to draw* the plane sections of a given cone. We give the details in Chapter 8, p. 214. It has been thought that Dürer's method was completely original, but we know that the works of Apollonius had just been revived from classical sources, and I feel that Dürer's construction is entirely in the spirit of Apollonius.

Dürer's construction of an ellipse is shown in Figure 14, and will be explained in Chapter 8. The ellipse shown is egg-shaped, whereas a true ellipse has *two* axes of symmetry. It has been thought that Dürer, like any schoolboy, felt that an ellipse was bound to widen with the widening of the cone of which it is a

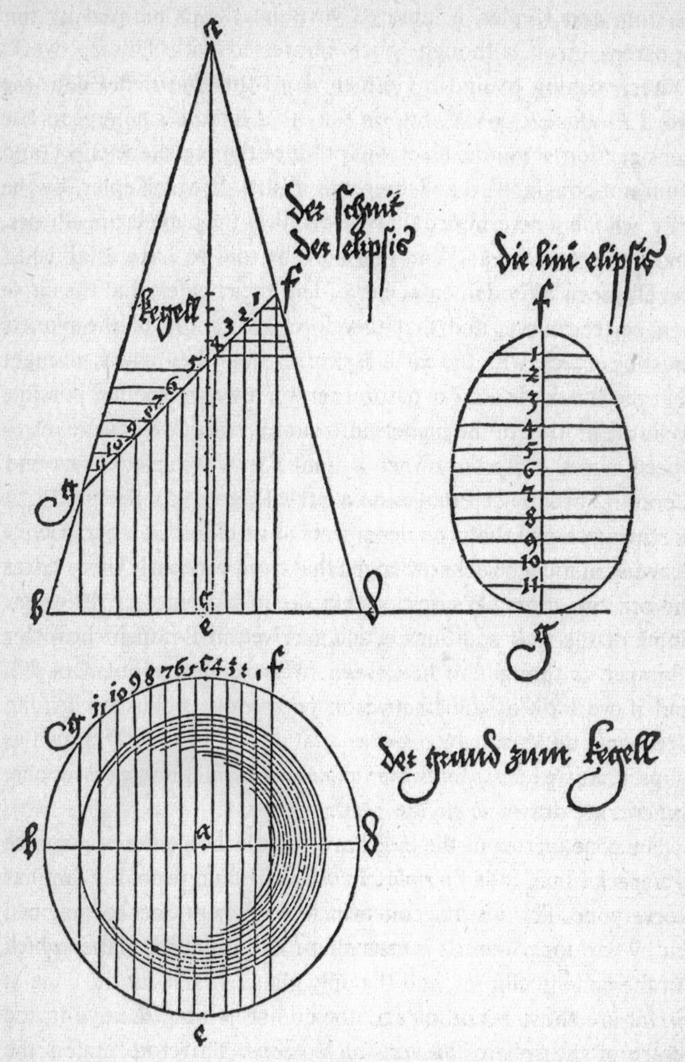

der schnit
der elipsis
die lini elipsis
kegell
der grund zum kegell

Figure 14

section, and Kepler, a hundred years later, was amused by this apparent error, although much impressed with Dürer's work. Dürer, wishing to find a German word for *elipsis*, did coin the word *Eierlinie*, egg-line, but on the other hand if one repeats the construction, a minute error will produce the egg-shape effect, and I am not convinced that Dürer was misled. It was Kepler, by the way, who first determined that the orbits of the planets are ellipses, and not epicycles, and can therefore be said to have established the ellipse as a fundamental curve. The ancients felt that the circle is a perfect curve, and that therefore the motion of the planets must be circles, with the earth as centre. But when it was thought that the sun made a more natural centre, it was not found possible to fit the motion of the planets into circles, so *epicycles* were introduced, curves traced out when a point P moves in a circle around a point Q which itself moves on a circle (Figure 15). See p. 218.*

Having shown that a circle appears as an ellipse in a perspective drawing, it must be acknowledged that from very early times it was the practice to depict circles as ellipses in paintings or drawings. Some of the wall paintings which survived in Pompeii show this (Pompeii is thought to have been overwhelmed about A.D. 79), and if we look at the treatise on perspective published by Jan Vredeman de Vris in 1604 we see that circles are clearly drawn as ellipses. But when *spheres* are delineated, something goes wrong. Spheres are drawn as *circles*.

The cone, vertex at the eye, drawn to all the points of a sphere is a special one, called a *right circular cone*, and probably familiar to everyone. This was the cone which the ancient Greeks imagined cut by various planes. It is clear that there are many planes which cut the cone in ellipses, and the only planes which cut the cone in *circles* are those perpendicular to the line joining the eye to the centre of the sphere, *the axis of the cone*. Therefore, unless the

*Some *epicycloids*, curves traced out by a point on the circumference of a circle which *rolls* on the outside of a fixed circle, are shown in Figure 101 and Figure 107. These are epicycles with, respectively, one and two cusps.

artist stands in a very special place, he should represent a sphere on his canvas as an ellipse, not as a circle.

M. H. Pirenne, in his fascinating book *Optics, Painting and Photography* (Cambridge University Press, 1970) gives some ex-

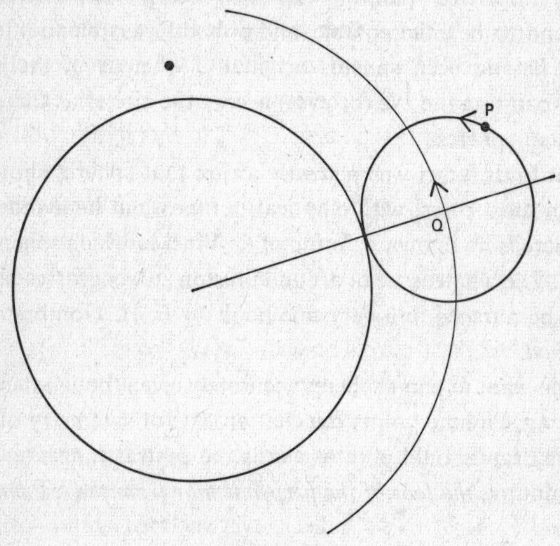

Figure 15

cellent pinhole photographs of spheres (Plate 15). A pinhole camera photograph reproduces exactly what we are after when we draw or paint a scene. As we can see, the sphere on the roof of the Church of St Ignazio, in Rome, appears in perspective as an ellipse. Pirenne has also photographed five spheres resting on top of cylinders, and this pinhole photograph, for which we must refer to his book, shows very clearly how the perspective drawing of a sphere can vary from a circle to an ellipse. But artists, for centuries, have insisted on drawing circles, not ellipses.

We now enter a thorny region of controversy. Raphael painted *The School of Athens*, now in the Vatican, with spheres represented by circles. But his painting is by no means an example of uniocular perspective. In fact there seem to be several points from which the picture may be viewed. Also, an experiment was conducted in the nineteenth century in which the 'incorrect' circles in the Raphael painting were replaced by 'correct' ellipses, and the resulting 'improved' painting was shown to people. The ellipses were found to be unacceptable, and nobody was able to interpret them as having been spheres originally, whereas in the actual Raphael painting the eye (or eyes) accept the circles as the representation of spheres.

Leonardo da Vinci was perfectly aware that spheres should be drawn, in most cases, with elliptical outlines, but he avoided the issue. There is no known painting of da Vinci in which spheres are depicted. The whole issue of art and illusion still engrosses people, and can be pursued in a very fine book by E. H. Gombrich, *Art and Illusion*.

If we do want to paint spheres accurately, even though they may look wrong, Pirenne points out that an axis of symmetry of each ellipse we draw should point towards the central vanishing point of the painting, *the foot of the perpendicular from the eye onto the canvas*.

This looks rather a hard theorem, but let us investigate it. We consider a right circular cone formed by joining the eye V to the points of a sphere. Let the plane of our canvas be called α. Then the plane cuts the surface of the right circular cone, since our eye is at V, in the curve which represents the sphere. We suppose that this curve is at a finite distance from V, so that it is an ellipse. If any of its points wandered off to infinity, it would be a parabola or a hyperbola.

Let N be the foot of the perpendicular from V onto the plane α (Figure 16). Through the line VN we can draw an infinity of planes. Consider the one which passes through the centre of the given sphere, and call it β. Then this plane β, since it passes through VN, which is perpendicular to α, is also necessarily perpendicular to α. Let the intersection of the planes α and β be the line n. This line

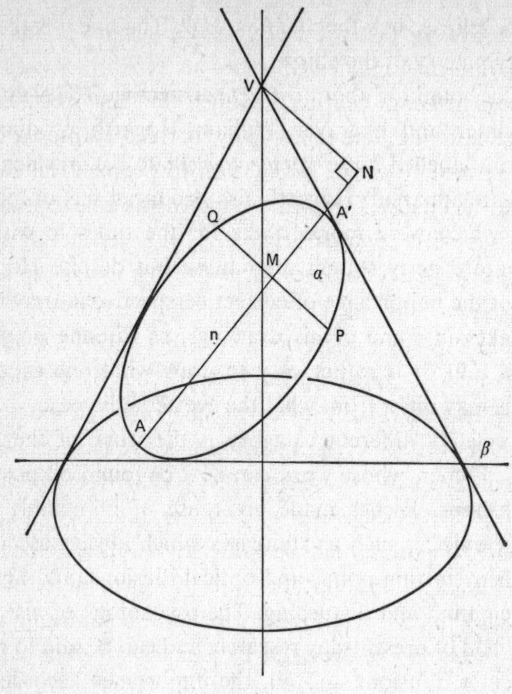

Figure 16

passes through N, since both α and β pass through N. It is asserted that the line n is an axis of symmetry for the ellipse.

This is readily seen if we take the plane of our paper for the plane β, which contains the axis of our right circular cone. The line n is the line $AA'N$, where A and A' are points on the ellipse, the plane of the ellipse being perpendicular to the plane of our paper. The surface of the cone has the plane $VAA' = \beta$ as a plane of symmetry, since any plane through the axis is a plane of symmetry. If from any point P on the ellipse we drop a perpendicular PM onto the line $n = AA'$, this line extended will meet the surface of the cone again in a point Q, where $PM = MQ$. But the line PMQ lies in the plane of the ellipse, and therefore Q is a point on the ellipse such that $PM = MQ$. The line n is therefore an axis of symmetry of the ellipse.

We have been talking about *correct* perspective. The eighteenth-century painter and engraver William Hogarth produced an amusing print labelled *False Perspective* (Plate 16) in which many absurdities are apparent. Hogarth also produced sets of 'correct' engravings which were moral tracts for the times in which he lived, and are eagerly sought after now. But despite Hogarth's awareness of the importance of correct perspective in drawing, he made mistakes in some of his drawings, as Pirenne points out (Pirenne, p. 169). It is rather easy to draw what one expects to see, and this may differ from what the eye actually sees.

In a completely different category is the work of the Dutch artist M. C. Escher, whose work can now be found on posters in many bookshops. Escher made great use of impossible three-dimensional objects, such as staircases which always ascend but return to their starting point, and optical illusions, and his work is both fascinating and disturbing. The psychology of vision is a very active field of present-day research, and can be said to restore to the observer, illusions and all, the importance accorded him during the Renaissance.*

As we have stressed, the position of the eye when viewing a work carried out strictly according to the theory of uniocular perspective is of the utmost importance. Pirenne gives some excellent photographs of a famous *tour-de-force* achieved by Fra Andrea Pozzo, a Jesuit monk who lived from 1642 to 1709. This is a painting on a hemi-cylindrical ceiling in the church of St Ignazio, in Rome. The painting represents St Ignatius being received into Heaven, and Pozzo wrote a book explaining how he did it.

The spot on the church floor where the one-eyed spectator is supposed to stand so that he can get a correct view is very plainly marked, and when he does stand there, the painting seems to reach

*See *Eye and Brain, The Psychology of Seeing*, by R. L. Gregory.

up far beyond the church ceiling, with columns stretching right up into Heaven. But if the painting is viewed from almost any other point on the church floor, the columns look bent, and the total effect is ludicrous.

Even with an ordinary painting exhibited on a vertical wall, it can be extraordinarily difficult to find the spot where the painter stood. There is a lovely painting in the National Gallery in London by Hans Holbein, called *The French Ambassadors* (Plate 17), in which many contemporary mathematical and astronomical instruments are displayed, apparently part of the furnishings of ambassadors in the early sixteenth century. In the foreground of the picture there is a curious elongated distortion, an anamorphosis of a human skull, which can be seen undistorted, that is as one expects a skull to look, if one stands at the right spot in front of the picture. So far, I have been unable to discover the right spot, and on one occasion my twistings and turnings were viewed with unsympathetic concern by one of the Gallery attendants, and I thought it was better to desist!*

Our last words on perspective in this chapter are a reference to the fact that we have only dealt with European perspective. There are paintings on the walls of caves in Ajanta, India, dating from the sixth century, which show that several vanishing points were used for the same system of parallel lines. There is a vanishing line, but sometimes there is more than one! Even more remarkable are the Chinese brush paintings created some time between the tenth and the thirteenth century, which show a vanishing line placed *behind* the spectator.

We have strayed quite a way from Dürer's *Underweysung*, and we now return to the Second Book, where he emphasizes the construction of regular polygons, especially those which do not derive

*In 1974 the National Gallery put this painting on special exhibition. The frame was removed and the painting was hung much lower than usual. It is now possible to see the skull very clearly without endangering oneself physically or socially.

immediately from the square or the equilateral triangle. The only techniques we have discussed, so far, are in connection with the bisection of a given circular arc, which enables us to construct a right angle, and the construction of an equilateral triangle, which gives us an angle of 60°. This angle can also be constructed by stepping round a circle, using a modern compass (Figure 17). We

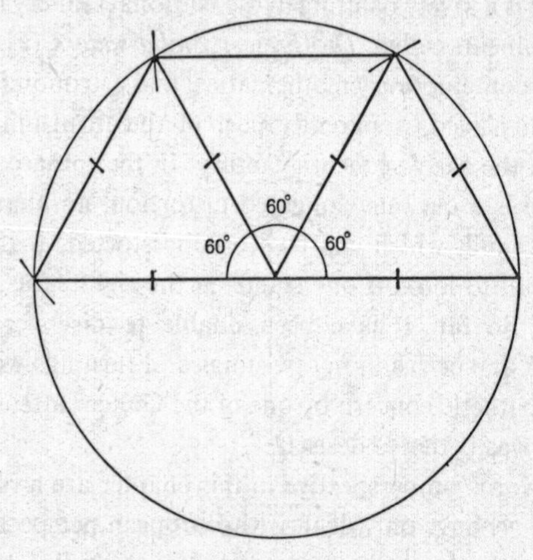

Figure 17

also showed how to find one fifteenth of a circumference, this having been mentioned by Vitruvius in an astronomical connection (p. 39).

Dürer's constructions are not all original, and some can be found in the *Geometria deutsch*, a compilation which appeared about 1484. This contains an approximate construction for a regular pentagon with a rusty compass which is still the quickest method known for practical drawing.

Let AB be the fixed stretch of the compasses, and draw the circles centre A and centre B, each circle having radius AB. Let the circles intersect at C and D (Figure 18). Then $AB = AD = BD$, as we already know. Draw the circle centre D and radius DA.

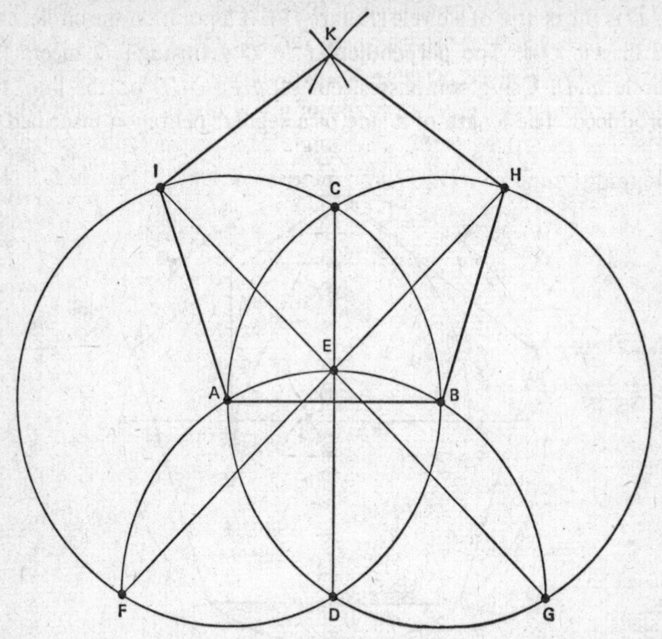

Figure 18

This will pass through B, and let it cut the line CD in E, and the circles centres A and B respectively in F and G. Produce the line FE to cut the circle centre B in H, and produce GE to cut the circle centre A in I. Then the intersection K of circles centres H and I respectively give the fifth vertex of the regular pentagon.

Dürer does not emphasize in this case that the construction is an approximate one. In fact it can be shown, using trigonometry, that the angle ABH is $108° \, 21' \, 58''$, instead of 108°. The angle BAI = angle ABH, and each of the angles BHK, AIK is slightly larger than 107°, and angle HKI is slightly larger than 109°, and this would hardly be detectable in a drawing.

Dürer also gives a theoretically exact construction for a regular pentagon, taken from Ptolemy's great work on astronomy, the *Almagest*.

O is the centre of a circle (Figure 19), A a point on the circle, and E bisects OA. The perpendicular to OA through O meets the circle in D. Using compasses, cut off $EF = ED$ on the line AO produced. The length of a side of a regular pentagon inscribed in

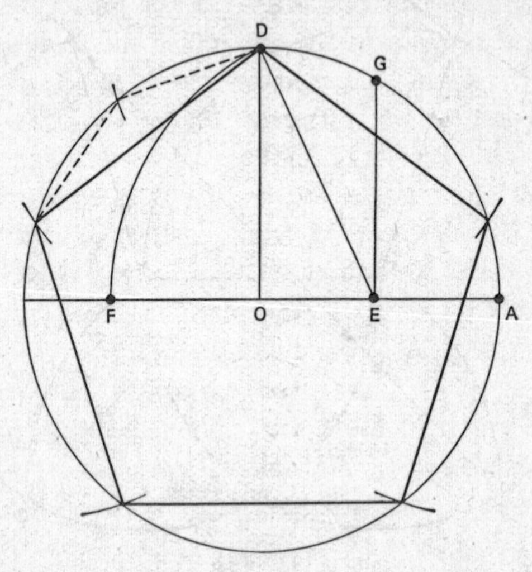

Figure 19

the circle is equal to the length DF. Moreover, OF is a side of a regular decagon (10-gon) inscribed in the circle, and if EG be perpendicular to OA, Dürer states that EG is *an approximation* to the side of a regular heptagon (7-gon) inscribed in the circle. The proofs of some of these statements are suggested in the Exercises at the end of this chapter.

Dürer's interest in the construction of regular polygons is a reflection of their use in the Middle Ages in both Islamic and Gothic decoration and, after the invention of firearms, in the construction of fortifications. Curiously enough, very few buildings have ever been constructed with a regular pentagon as a ground-plan. The Pentagon, in Washington, D.C., is an outstanding exception.

The medieval constructions of regular polygons were, of course, approximate, but they were, and had to be, simple, preferably not even calling for a change in the opening of the compasses, which had no device to recapture an opening, once it was changed. Leonardo da Vinci also wrote a lot on polygons, but it was Dürer, rather than Leonardo, who transmitted the medieval constructions to the future. Dürer knew Euclid, but he does not give Euclid's construction for a regular pentagon, which is a theoretically exact one, as are all Euclid's constructions. Euclid makes no attempt to trisect a given arc, and Dürer knew that this was impossible, although a proof, which is algebraic, did not appear until the nineteenth century. In fact it was Gauss who first discovered which regular polygons can be constructed with ruler and compasses, and this will be discussed in Chapter 9, p. 257. Dürer does give *approximate* constructions for the trisection of an arc. These are the kind which are continually being discovered by *angle-trisectors*, enthusiastic amateurs of geometry who refuse to believe the proof that an angle *cannot* be trisected with ruler and compasses *exactly*.

Euclid's construction of a regular pentagon involves the division of a line *in extreme and mean ratio*, a ratio which was later called *the golden ratio*, and which proved to be of enormous interest to artists and architects for centuries, as we shall see in Chapter 4. Dürer does not mention this ratio, although he must have known Fra Luca Pacioli's ecstatic book on the *Divina Proportione*. A line ABE is said to be divided at B in the *golden* ratio (or *divine* ratio) if the ratio of the larger part to the smaller is equal to the ratio of the whole to the larger part.

As an equation, this is $AB/BE = AE/AB$. If we write $AB = a$, and $BE = a/\mu$, so that the golden ratio is $AB/BE = \mu$, we have the equation:

$$\mu = 1+1/\mu,$$

so that μ satisfies the quadratic equation

$$\mu^2-\mu-1 = 0.$$

This has one positive root,

$$\mu = (\sqrt{5}+1)/2 = 1{\cdot}618\ldots$$

We note that $1/\mu = (\sqrt{5}-1)/2$, since $(\sqrt{5}+1)(\sqrt{5}-1) = 5-1 = 4$.

If we construct a square with side AB (Figure 20), bisect AB at

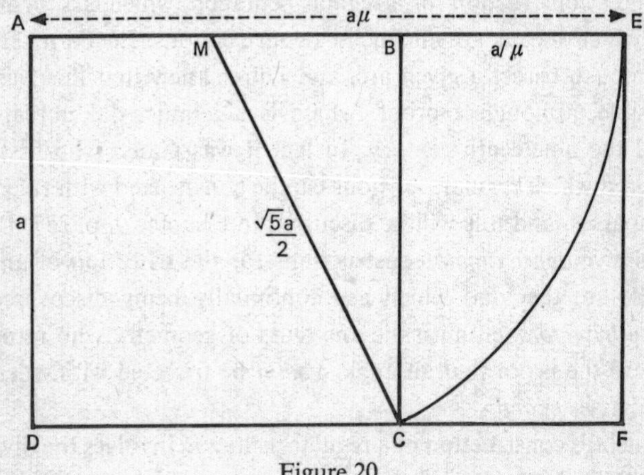

Figure 20

M, and with radius MC draw an arc of a circle centre M to cut AB produced at E, then AE is divided in the divine ratio at B.

To prove this, we note that, by the Pythagoras Theorem,

$$(MC)^2 = a^2+(a/2)^2 = 5a^2/4,$$

and therefore

$$AE = a/2+ME = (\sqrt{5}+1)a/2 = \mu AB.$$

The rectangle $AEFD$, having the side $AE = \mu AD$, is called *a golden rectangle*. $ABCD$ is a square, and we observe that the rectangle $BEFC$ is also a golden rectangle, since $BC = a = \mu BE$. This at once suggests further divisions of the rectangle $BECF$, and in fact we shall consider this in Chapter 4.

Let us see how Euclid uses the golden ratio to construct an angle of 72°, which is the angle subtended by a side of a regular pentagon at the centre of its circumscribing circle. Begin with a line ABE divided in the golden ratio at B (Figure 21). With com-

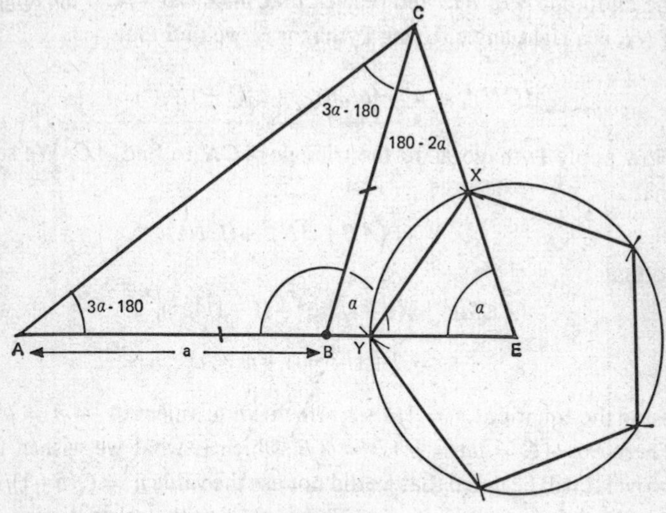

Figure 21

passes centred at B and at E, and radius AB in each case, draw two arcs of circles to intersect at the point C. We shall show in a moment that $AC = AE$.

Assume this and let the equal angles EBC and CEB be called α. Then since $AC = AE$, the angle ACE is also α. Using the theorem that the angles in a triangle add up to 180°, the angle $BCE = 180-2\alpha$, and the angle EAC is therefore equal to $3\alpha-180$. The angle $ABC = 180-\alpha$, and so, adding the angles of triangle ABC, $180 = (3\alpha-180)+(3\alpha-180)+(180-\alpha)$. Therefore $5\alpha = 360°$ and $\alpha = 72°$.

The triangle BEC is now such that its base angles are each twice as large as the remaining angle of 36°. To obtain a regular pentagon, all we have to do now is to construct any circle centre

E, cutting EC in X and EB in Y, and then XY is a side of a regular pentagon inscribed in the circle, and we can step round the circle and obtain the other sides.

For the proof that $AC = AE$, we suppose that C is joined to the midpoint N of BE, and remark that since $CB = CE$ the angle CNE is a right angle. Using Pythagoras, we find that

$$(CN)^2 = a^2-(a/2\mu)^2 = a^2(1-1/4\mu^2).$$

Now apply Pythagoras to the triangle ACN to find AC. We see that

$$(AC)^2 = (AB+BN)^2+(CN)^2$$

so that

$$(AC/a)^2 = (1+1/2\mu)^2+(1-1/4\mu^2),$$

$$= 2+1/\mu = 1+\mu = \mu^2,$$

using the equation: $\mu = 1/\mu+1$, which is equivalent to $1+\mu = \mu^2$. Therefore $AC = \mu a = \mu AB = AE$, which is what we wished to prove. It will be noted that we did not use the value $\mu = (\sqrt{5}+1)/2$, but the defining property of μ. This simplified the calculations.

Kepler (1571–1630) said, of the golden ratio: *Geometry has two great treasures: one is the Theorem of Pythagoras: the other the division of a line into extreme and mean ratio. The first we may compare to a measure of gold; the second we may name a precious jewel.*

It is difficult to recapture the excitement felt by Renaissance man when the beauties of geometry unfolded before his eyes, but one must report it. A late sixteenth-century Italian geometer, Pietro Antonio Cataldi wrote a whole monograph on Albrecht Dürer's construction of the regular pentagon, and it is known that this construction stimulated the imaginations of men like Cardano, Tartaglia and Galileo. We shall see (p. 92) that at one period of his life Leonardo da Vinci was so obsessed with geometry that he did no painting. We must not leave regular polygons without showing the lovely tiling pattern produced by Dürer, using the regular pentagon (Figure 22). The reader is invited to indulge in a similar activity in Exercise 5 at the end of this chapter.

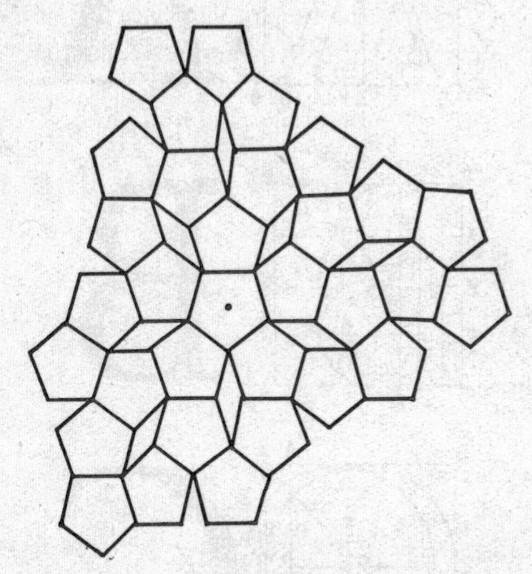

Figure 22

The Third Book of the *Underweysung* is of a purely practical character, and intended to illustrate the application of geometry to the concrete tasks of architecture, engineering, decoration and typography. Dürer was an admiring student of Vitruvius, but Dürer's designs break away from the classical tradition, as exemplified by the Roman architect. Dürer acquaints the Northern countries with another *secret*, that is *unpublished* work of the Renaissance, the geometrical construction of Roman letters (Figure 23 gives an example of Dürer's method). Here Dürer acts as a middleman. Similar constructions can be found in Fra Luca Pacioli's *De Divina Proportione*. It may be wondered why exact geometrical constructions are necessary. Letters can be drawn freehand fairly accurately when they are small, but larger

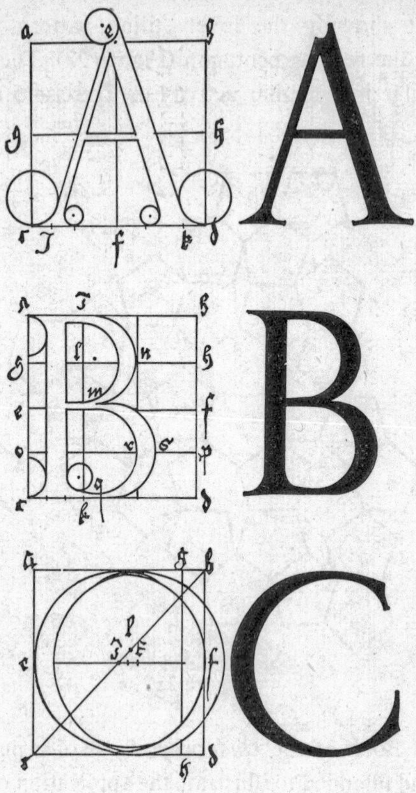

Figure 23

letters to be used on tombstones or buildings look wrong unless they are designed geometrically. When Dürer comes to Gothic letters he constructs these according to an original principle (Figure 24). He dispenses with circular arcs altogether, and instead of inscribing the letter in a large square, he builds it up from a number of geometrical units, such as squares, triangles or *trapezoids*, four-sided figures with a pair of parallel sides.

The Fourth Book of the *Underweysung* resumes the thread of the Second Book, and deals with the geometry of three-dimen-

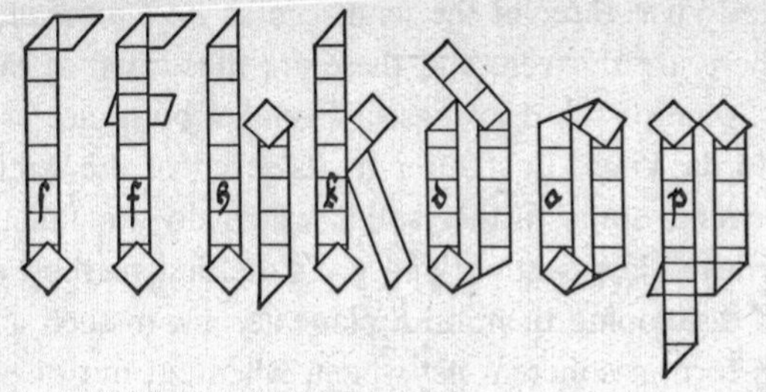

Figure 24

sional bodies, a field entirely disregarded, as far as we know, during the Middle Ages. At the beginning, Dürer discusses the five regular *Platonic* solids, so-called because they are briefly discussed by Plato (Figure 25). We shall deal with them in more

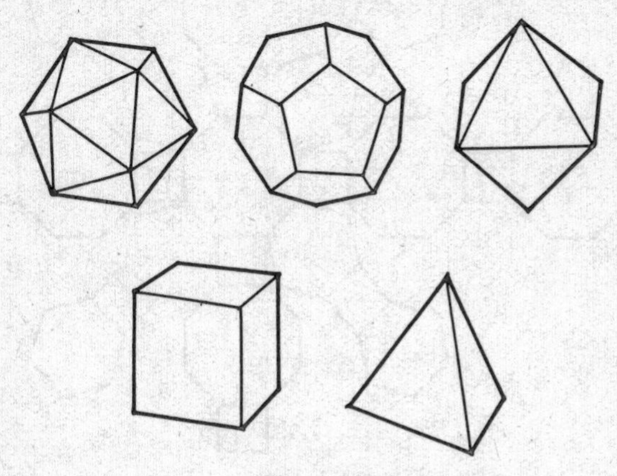

Figure 25

detail in Chapter 9. These solids had come into the limelight because of the revival, with the discovery of manuscripts, of Platonic studies, and because these surfaces were excellent models for use in perspective studies.

We have mentioned the *Divina Proportione* of Fra Luca Pacioli a number of times (p. 69). Pacioli discusses the Platonic solids, and gives three of the semi-regular *Archimedean* solids, of which there are thirteen, and these are illustrated in perspective images. Some of the drawings in Pacioli's book are ascribed to Leonardo da Vinci. But Dürer treats seven of the Archimedean solids, invents some further solids which do not fall under the Archimedean classification, and gives the first method ever published of developing them on a plane surface in such a way that the faces form a coherent *net* which, when cut out of paper and properly stuck together, will form an actual three-dimensional model of the solid in question (Figure 26). These *nets* will appear again (p. 285).

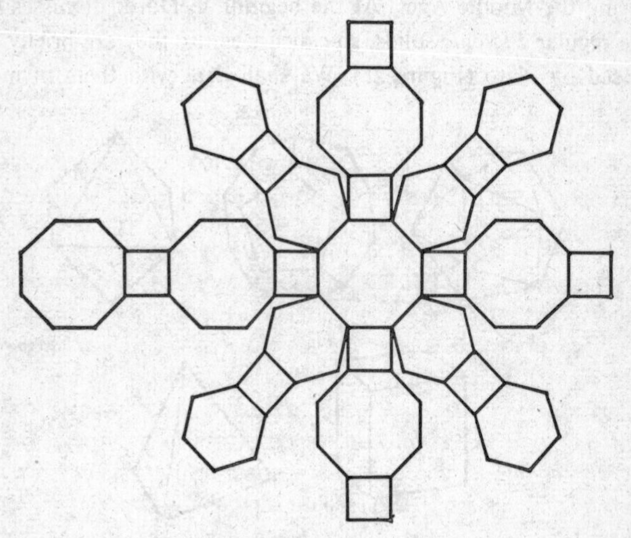

Figure 26

The section on solids is followed by a discussion of the *Delian problem*, the problem of doubling a cube (p. 31). Dürer gives three solutions to the problem, and we shall investigate some solutions in Chapter 8. This section on the Delian problem leads to the final section of the *Underweysung*, the chapter on perspective, which shows how to construct the perspective drawing of a lighted cube placed on a horizontal plane. To Dürer, as we have already remarked, and to many of his contemporaries, perspective meant the crown and keystone of the majestic edifice called Geometry.

We conclude our discussion of Albrecht Dürer with a brief look at his *Four Books of Human Proportion*, 'conceived and written by Albrecht Dürer of Nuremberg for the use of all such as bear a love towards this Art'.

They were printed posthumously in 1528, and seen through the press by Dürer's great friend Pirkheimer. The influence of Vitruvius is clear. To quote Panofsky, the figures are constructed thus: '. . . the length of the head is one eighth of the total height, that of the face (divided into three equal parts, the brow, the nose and the rest) one tenth, and the width of the breast from shoulder to shoulder one quarter . . .',* and so on.

The figures themselves are constructed with a great use of geometry over and above the use of ratios (Plate 18). To quote Panofsky again:

> The total length and general axis of a standing body is determined by a basic vertical which runs from the heel of the standing leg to the top of the head, and goes through the pit of the stomach. The pelvis is inscribed in a trapezoid, and the thorax in a square (in some female figures in a vertical rectangle), and the axes of these, meeting at the pit of the stomach, are slightly shifted against the basic vertical. The knee of the standing leg and thereby the length of both thighs is found by bisecting the line which connects the hip-point with the lower terminal of the basic vertical. The head, if turned in full profile, is inscribed in a square, and the contours of the shoulders, hips and loins . . . are determined by circular arcs.

Dürer's original scheme 'did violence, in some measure to nature. It denied individual differences and hardened into geometrical curves what should be an organic undulation. At a com-

*Panofsky, *The Life and Art of Albrecht Dürer*, Princeton University Press, 1943, p. 262.

paratively early date the number of circular arcs determining the contours was restricted to three . . .',* and abandoned altogether in some drawings.

Later on, Dürer was to state explicitly that '*the boundary lines of a human figure cannot be drawn with compasses and ruler*'.

The *Vier Buecher* show very definite changes in Dürer's evolving concepts of the human figure. Dürer's visit to Italy and probable encounter with Leonardo da Vinci mark a turning point in his career as a theorist of human proportions. Dürer investigated the measurements, in the course of his studies, of several hundred living persons, but never pursued any anatomical studies. Leonardo, on the other hand, was one of the first artists to study anatomy. The feeling that there is much more to the human figure than can be realized by overall measurements may have helped to modify Dürer's ideas. He describes five types of human physique and character in the First Book, and although he adds eight further types in his Second Book, the idea of a *rule* has been abandoned, and the influence of Vitruvius no longer manifests itself in actual measurements, but merely in the fact that some of the figures are inscribed in a square or a circle.

The Third Book gives various methods which enable an artist to change the proportions of any figure described in the first two Books in any way, yet on the basis of a consistent geometrical principle. These methods consist of different kinds of geometrical mappings, and some are *affine* maps, which we shall describe. The effect of these maps is to produce *caricatures* (Plate 19), and here the influence of da Vinci may show itself again, since da Vinci's Notebooks contain many such drawings. Da Vinci may have just sketched real people, but Dürer showed how such distortions can be arrived at geometrically, perhaps another indication that *God geometrizes*!

An affine map is a one-to-one map in which points map into

* Panofsky, *The Life and Art of Albrecht Dürer*, Princeton University Press, 1943, p. 264.

points, lines map into lines, and *the line at infinity* in the one plane maps into *the line at infinity* in the plane of the map. More simply, *parallel lines* in the one plane map into *parallel lines* in the plane of the map.

In our perspective mapping (Figure 12) this would mean that the eye V is at infinity, or that the planes α and α' are parallel to each other. In the second case we do not obtain a distorted picture of what the eye sees, merely an enlargement or reduction of the view, but if we imagine the eye V to be the sun, or a bright light a long way off, it is clear that we can hold a transparency so that the projection on the ground is a distorted image of the picture shown by the transparency. Although a distortion, lines which are parallel project into parallel lines, and if three points A, B, C are collinear in the transparency, and project into the three collinear points A', B' and C', it can be shown that the ratios AB/BC and $A'B'/B'C'$ are equal. Use can be made of this fact in sketching an affine map of a given picture, as is suggested in Exercise 4 at the end of this chapter.

We have produced ample evidence that Dürer was a natural geometer. The *Vier Buecher* – translated into Latin as early as 1532, and then into many European languages – laid the foundations of scientific anthropometry. It must be pointed out that Michelangelo and a host of other Italian artists and theorists called the Books a futile enterprise and a waste of time. But Dürer hardly intended the artist to do any more with the help of the Books than to obtain a firm foundation. Dürer says: 'It is not my opinion that an artist has to measure his figures all the time. If thou hast learned the art of measurement and thus acquired theory and practise together . . . then it is not always necessary to measure everything all the time, for thy acquired art endows thee with a correct eye.'

There is much more than we could say about Dürer's attitude towards art, but we must conclude this chapter. As a final word, Renaissance artists, like those of the present day, were always try-ing to justify themselves, and Dürer, more than any other artist of his period, can be held responsible for heightening the self-consciousness of the modern artist. Dürer was one of the first artists to make notes on his prints and drawings, so that he could recall the subject of each work and the conditions under which it was completed. He also collected the works of other artists, and made notes on their work. So when you enter a gallery and see a few objects on the walls, with a centrepiece consisting of a long discourse by the artist with a description of his (or her) emotions before, during and after the act of creation, plus a philosophy of the times which combines his (or her) attitudes towards civil liberties, pollution, abortion, organic food and whatnot, perhaps Dürer can be held to have been partly responsible.

Exercises

1. Draw a square $ABCD$ divided into 16 squares (Figure 13), and construct a perspective drawing in which AB and CD are parallel to the horizontal in the picture plane. Draw the equidistant transversals in your picture.

2. Inscribe a circle in the square $ABCD$ of Exercise 1 above, and obtain 12 points on the perspective map of the circle, as described on p. 56. If you have the patience, obtain a finer mesh in the original square $ABCD$ by subdivision of each of the 16 small squares into 4 squares, and hence a more accurate representation of the ellipse inscribed in $ABC'D'$.

3. Make a perspective drawing in which the vanishing line is *behind* the viewer, and another in which there is more than one vanishing line.

4. Construct a square $ABCD$ as in Exercise 1, divided into 16 smaller squares, and draw an object in the square, a flower, a face or a vase, observing the relation of your drawing to the 16 squares of the mesh. Now draw a parallelogram $A'B'C'D'$ in which the sides $A'B'$ and $B'C'$ are not necessarily equal, and subdivide it into 16 equal parallelograms. Draw the transform of your original figure, assuming that $ABCD$ is mapped onto $A'B'C'D'$, and that the ratios in which points of your figure divide segments parallel either to AB or to BC are unchanged by

the transformation. (If P is a point in your original figure, and a line through P parallel to AB meets AD in E and BC in F, then if P' is the corresponding point in the mapped figure, and a line through P' parallel to $A'B'$ meets $A'D'$ in E' and $B'C'$ in F', we are assuming that $EP/PF = E'P'/P'F'$. With the help of another line drawn through P parallel to BC this time, we can locate P' exactly. Of course, all that is needed here is an approximate construction, as in the affine map in Plate 19.

5. Cut a regular pentagon out of cardboard, and use it to make tiling patterns, as Dürer did (Figure 22). Can you discover any others?

6. Using an inexpensive copy of the St Jerome print (Plate 11), find the vanishing points of the various sets of lines which may be supposed parallel in the original scene. Noting that the chair is at 45° to the edge of the table, can you obtain an approximation to the distance of the artist from the original imagined scene?

7. Construct some capital letters geometrically as shown in Figure 24, or by using variations on the constructions shown there, and try some using parallelograms, and not circles.

8. In Figure 19, Dürer's construction of a regular pentagon, take $OA = 1$, and show that $(EF)^2 = 5/4$, and $(DF)^2 = (5-\sqrt{5})/2$. If you know some trigonometry, use the fact that $\sin 36° = \sqrt{10-2\sqrt{5}}/4$ to confirm the fact that $DF =$ a side of a regular pentagon inscribed in the circle, and also using the fact that $\sin 18° = (\sqrt{5}-1)/4$, confirm the statement that $OF =$ a side of a regular 10-gon inscribed in the circle.

9. In Figure 20, which shows a rectangle $AEFD$ in golden section constructed from a square $ABCD$, continue the construction by drawing a square $BEF'C'$, and show that EF is also divided in golden section at F'. Can you continue?

·3·

LEONARDO DA VINCI

LEONARDO DA VINCI (1452–1519) is one of history's enigmas. He has been hailed as a consummate artist, a genius of the first water, a great scientist who anticipated many inventions which were not realized until the twentieth century, and so on. Freud has written about him, as have many lesser men. There is no doubt that he was a great artist, and he was so regarded by his contemporaries, but he will always remain a mystery because he left no connected discourse comparable to the works of Vitruvius or Dürer, but only numerous works in manuscript, his *notes* on practically every subject under the sun.

These Notebooks were, for an inexplicably long time, unpublished, and indeed forgotten. On the other hand it is certain that in the sixteenth and seventeenth centuries their exceptional value was highly appreciated, and there was great competition for a few pages of manuscript. This explains why Leonardo manuscript can be found in royal libraries, in London, in Ashburnham Place, Sussex, in the Louvre, and of course in various parts of Italy, perhaps as widely scattered over Europe as any of the productions of Renaissance man.

How much time the various owners of Leonardo manuscripts spent in deciphering them is perhaps a question best not asked. 'He wrote backwards, in rude characters, and with the left hand,' says Vasari, the Italian biographer of Renaissance artists, and in fact Leonardo's writing is the most famous example of *mirror writing* in existence. Transcribers have always found that using a mirror for this writing does not help very much, and they have learned to read Leonardo without such aids. But of course, as with all manuscripts, there are other difficulties. Leonardo has a fashion of amalgamating several short words into one long one, and he does not use punctuation, accents, or any device which might make reading easier. A page may begin with some principles of astronomy, or the motion of the earth; then come the laws of sound, and finally some precepts on the subject of colour. Another page may begin with an investigation into the structure of the intestines, and end with philosophical remarks on the relation of poetry to painting. The pages were unbound and this made them easier to disperse, of course. Perhaps what made collectors so eager to possess even a few pages are the drawings. Most are in ink, but some are in red chalk, and some of the writing is in silverpoint.

The first entries date to about 1490, when Leonardo was already a mature person, and they continue until his death, a span of about thirty years. Curiously enough, the writing does not change throughout.

Our first quotation is: *Let no man who is not a mathematician read the elements of my work*

With this caution ringing in our ears, let us see what Leonardo says about perspective. We remember that the writers of the Renaissance did not regard perspective and optics as distinct sciences. Perspective, indeed, in its widest application, is the science of seeing. Leonardo says:

Among all the studies of natural causes and reasons Light chiefly delights the beholder: and among the great features of Mathematics, the certainty of its demonstrations is what pre-eminently tends to elevate the mind of the investigator. Perspective must therefore be preferred to all the discourses and systems of human learning. In this branch of science the beam of light is explained on those methods of demonstration which form the glory not so much of mathematics as of physics, and are graced with the flowers of both.

Leonardo divides Perspective into three divisions. The first includes the diminution of size of opaque objects, the second treats of the diminution and loss of outline in such opaque objects, and the third treats of the diminution and loss of colour at long distances.

All the problems of Perspective are made clear by the five terms of Mathematics which are the point, the line, the angle, the surface and the solid. A point is not part of a line. The smallest natural point is larger than all mathematical points, and this is proved because the natural point has continuity, and anything that is continuous is infinitely divisible: but the mathematical point is indivisible because it has no size . . . If one single point placed in a circle may be the starting point of an infinite number of lines, and the termination of an infinite number of lines, there must be an infinite number of points separate from this point, and these, when reunited, become one again, whence it follows that the part may be equal to the whole.

In this extract from the Notebooks it is clear that Leonardo is discussing Euclid. One of the *Common Notions* in Euclid reads: *The Whole is Greater than Any One of its Parts*, and although this is acceptable when we are dealing with a finite collection of ob-

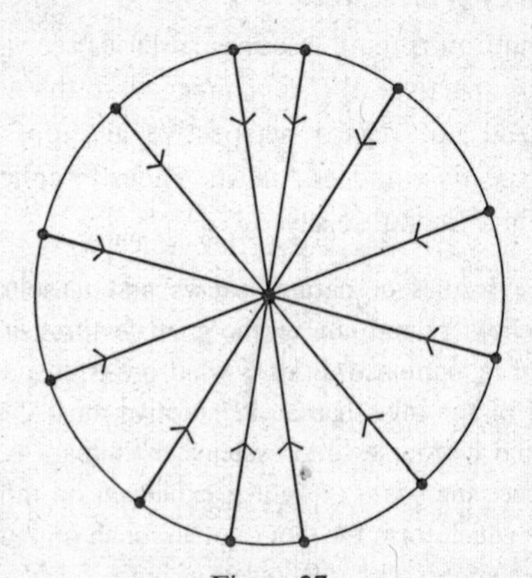

Figure 27

jects, it is no longer true when we consider *infinite* sets. Leonardo has a very clear idea of the difference between a mathematical point and an actual, physical point. Euclid defines a point as follows: *A Point is that which hath no Part*, but makes no use of this definition subsequently.

Leonardo is thinking of an infinite set of lines passing through a point (Figure 27), of taking a point on each line near the central point, and then allowing all these points to move in towards the central point. A hundred years later Galileo (1564–1642) was still worrying about the problems of infinity and Euclid's Common Notion, remarking that every integer n can be paired with the even integer $2n$, so that the pairing looks like this:

$$\begin{array}{cccccccccc} 1 & 2 & 3 & 4 & 5 & 6 & 7 & \ldots & n & \ldots \\ \updownarrow & \updownarrow & \updownarrow & \updownarrow & \updownarrow & \updownarrow & \updownarrow & & \updownarrow & \\ 2 & 4 & 6 & 8 & 10 & 12 & 14 & \ldots & 2n & \ldots \end{array}$$

which indicates that *there are as many even integers as integers*, but there are obviously more integers than just even integers. Real progress was not made until the nineteenth century, when Cantor began to think of the problem.*

Leonardo discusses the other fundamental concepts of geometry, and when he comes to boundaries, says: 'The boundaries of bodies are the least of things . . . wherefore O painter! do not surround your bodies with lines.' Four hundred years later Van Gogh shocked the artistic world by painting very visible lines round his sun-flowers and chairs.

Leonardo is very precise on the concept of the pyramid of sight, and he must have obtained a very satisfactory verification of his ideas from the *camera obscura*. He is the first writer to discuss this device, which is exactly like a pinhole camera, but without photographic film at the back. Instead of film, a sheet of ground glass fills the back of the box, the pinhole being at the front, and pointed towards a scene. This appears, in natural

*Pedoe, *The Gentle Art of Mathematics*, chapter 3.

colours on the screen, but upside-down. With the help of a lens, the scene can be viewed the right way up. Here is a ready-to-hand way of depicting a landscape on a flat surface, and there is plenty of evidence that landscape painters made use of the *camera obscura*. Stately mansions used to have a special room fitted as a *camera obscura*, so that visitors could survey the surrounding countryside with its help. Many such simple pleasures have been forgotten.

With the help of the *camera obscura* the existence of *vanishing points* is easily demonstrated, but Leonardo has a simpler proof to convince doubters. 'When you walk by a ploughed field, look at the straight furrows which come down with their ends to the path where you are walking, and you will see that each pair of furrows will look as though they tried to get nearer and meet at the farther end.'

Leonardo refers to an apparent paradox in the theory of perspective which is still of interest. It has also been discussed by Piero della Francesca. If you paint a row of columns, such as those in a temple façade, from a point in front of them, the columns on the side will come out wider than those in front, although they are farther away. The columns are all supposed to be of the same width, of course, and a section at right angles to the columns gives us a number of circles in line. It is sufficient to consider three of them (Figure 28), and as we see in the diagram, the images of the two outer circles in the picture plane are wider than the image of the nearer circle. That is, PQ is greater than RS.

Leonardo knew that the size of an object depends not at all on the distance from the eye, but on the angle subtended at the eye. Are we comparing objects of the same size? One might think that we are, since the circular sections are all of the same radius. But we are actually viewing more of the circumferences of the two outer circles than of that of the nearer circle, and yet the angle subtended at the eye is smaller. Think of a circle sliding towards you along a pair of embracing lines which emanate from the eye.

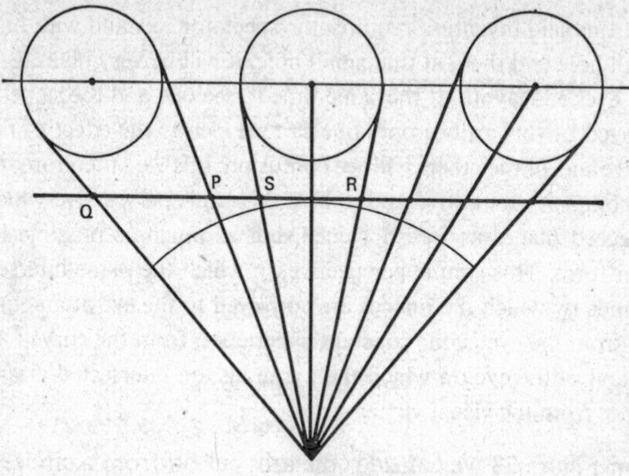

Figure 28

The lines have to open out as the circle approaches, and they close as the circle moves away. The greatest amount one can ever see of the circumference of a circle is one half, when the circle is a very long way away, and the smallest part one can ever see is as small as you may wish, when the circle is very near to the eye.

As we have already noted (p. 62), the accurate delineation of a scene may lead to mental doubts as to its verisimilitude. If you were in front of a very long wall, would you paint the top of the wall as a horizontal line running straight across your canvas? Theoretically, you should, but it is a matter of experience that if you look at the nearest point of the wall, the distant points of the wall, viewed from the side of your eye, look as if they are curving away from the wall, and towards you. Large pinhole camera photographs look unreal, because the eye does not take in too much of a scene at any given instant, and makes continuous and complex adjustments.

Leonardo was very aware of the difficulties caused by uniocular perspective, and in the same note in which he refers to the apparent paradox we have just described says, with reference to uniocular perspective:

But this said invention requires the spectator to stand with his eye at a small hole and then, at that small hole, it will be very plain. But since many eyes endeavour at the same time to see one and the same picture produced by this artifice only one can see clearly the effect of this perspective and all the others will see confusion. It is well therefore to avoid such complex perspective and to hold to simple perspective which does not regard planes as foreshortened, but as much as possible in their proper form. This simple perspective, in which the plane intersects the pyramids by which the images are conveyed to the eye at an equal distance from the eye is our constant experience, from the curved form of the pupil of the eye on which the pyramids are intersected at an equal distance from the visual virtue.

If in Figure 28 we consider the arcs cut off from a circle, centre at the eye, the farther circle cuts off a smaller arc than the nearer circle, and so, using Leonardo's *simple perspective* on a curved canvas, the viewer would not be disturbed.

We have already noted (p. 62) that Raphael painted spheres as the viewer expects to see them, rather than as the spectator with his eye at a little hole will see them, and that in the same painting there is more than one preferred viewpoint. Leonardo was an immensely practical artist, and not at all disposed to adhere rigidly to a theory if the resulting picture made the viewer uncomfortable. Of course, the possibilities of adjustment in the human eye are very great, so much so that people become rather disturbed if this is pointed out to them. We have all watched the full moon rising above the horizon. It looks enormous, but when it is at its highest point in the night sky, it looks much smaller.

Years ago I was having a meal in a guesthouse* with about a dozen people, of all professions, when this subject came up. I remarked that actually the moon was the same size all the time, and

*This guesthouse, Llethr, overlooking Pendine Sands in Carmarthenshire, run for many happy years by Arthur and Judy Giardelli, has just ceased to exist. Many people, including a recent Lord Chancellor, will always associate it with music, art, and long walks along peaceful beaches.

that it only looked smaller as it rose because, initially, there were other nearer objects to compare it with, such as trees, buildin s and so on, whereas once the moon was way up in the sky, it was all on its own. I knew that the distance of the moon from my eye could hardly change much while it was rising, and I also knew that therefore the angle subtended by the moon at my eye could not change much, and that its size depended upon this angle. One had to allow for the effects of atmospheric refraction, the strain on the eye muscles when the head is tilted, the consequent accommodation of the eye, and so on.

To my surprise, my point of view was hotly attacked by quite a number of the other men present, and quite soon, as in all disputes where one party is thought to have a superior attitude, there were thinly veiled hints that probably my morals, like my opinions, were also rather loose! Our host, an artist by profession, thought that he could resolve the dispute by phoning his father, a retired school teacher. The answer which was brought back to the dinner table only made matters worse. 'Tell your guests,' it said, without any preamble, 'that anything that Dr Pedoe says is sure to be right!'

Leonardo, whose opinions were always listened to with the greatest respect, was very interested in astronomy, the cause of eclipses of the moon, and so on, and makes it plain that he believed that the moon is subject to gravitational forces, just as things are on earth. He was well in advance of his contemporaries in his ideas on astronomy, and with his great interest in light he came near to anticipating Newton's discovery of the composite nature of light when he said '. . . white is not a colour, but the neutral recipient of every colour . . .' Elsewhere he discusses whether 'the colours of the rainbow are produced by the sun', and argues that they are not, since they occur in many ways without sunshine, as witness the effect if a piece of coarse glass with minute bubbles in it is held up to the eye. He then asks whether the eye has any part in producing the colours of the rainbow, and argues that it does not, by letting sunlight stream through the minute bubbles, and observing the rainbow effect produced on the ground. Newton performed the same experiment, but with a prism, and deduced the composite nature of white light.

As an artist, Leonardo was very interested in shadows, and his Notebooks contain many investigations into different kinds of shadows, and their interaction with each other. These continue to be of great interest to artists.

Leonardo's researches on the proportions and movements of the human figure seem to have been written up before the year 1498, since there is a reference to his work in a book published that year. But it seems that Leonardo did not consider his studies complete, and it is probable that the anatomical studies he pursued with so much zeal between 1510 and 1516 led him to reconsider the subject of proportion. Here are a few extracts from his Notebooks.

> Every man at three years old is half the full height he will grow to at last... Take a man three braccia (six feet) high, and measure him by the rule I will give you. The length of the hand is one third of a braccia, and this is found nine times in a man. And the face is the same, and from the pit of the throat to the shoulder, and from the shoulder to the nipple, and from one nipple to the other, and from each nipple to the pit of the throat...

There are pages of detail. He quotes Vitruvius the Architect as saying:

> If you open your legs so much as to decrease your height by one fourteenth, and spread and raise your arms till your middle fingers touch the level of the top of your head, you must know that the centre of the outspread limbs will be in the navel, and the space between the legs will be an equilateral triangle.

Leonardo's drawing (Plate 3) is reproduced in editions of Vitruvius which appeared in 1511 and 1521.

We have remarked that Dürer did not base his studies of the

human form on anatomical dissections, whereas Leonardo, far in advance of his time, was deeply interested. But he does issue this warning:

O, Anatomical Painter! Beware, lest the too strong indication of the bones, sinews and muscles be the cause of your becoming wooden in your painting ... It is indispensable to a Painter who would be thoroughly familiar with the limbs in all the positions and actions of which they are capable, in the nude, to know the anatomy of the sinews, bones, muscles and tendons so that, in their various movements and exertions he may know which nerve or muscle is the cause of each movement, and show only those as prominent and thickened, and not the others all over the limb, as many do who, to seem great draughtsmen, draw their nude figures looking like wood, devoid of grace so that you would think you were looking at a sack of walnuts rather than the human form, or a bundle of radishes rather than the muscles of figures.

As he seems to have anticipated so much, we must mention that Leonardo may also be held responsible for the discovery of what is called the Rorschach test nowadays. The subject is shown various inkblots in red and black, and asked to say what these suggest to him. From the answers, deductions are made about the subject's psychological state. Leonardo mentions:

... a new device for study, which although it may seem trivial and almost ludicrous, is nevertheless extremely useful in arousing the mind to various inventions. And this is, when you look at a wall spotted with stains, or with a mixture of stones, if you have to devise some scene, you may discover a resemblance to various landscapes, beautified with mountains, rivers, ... and these appear on such walls confusedly, like the sound of bells in whose jangle you may find any name or word you may choose to imagine.

Leonardo was one of the universal men of the Renaissance. He was an architect as well as a painter and sculptor. In 1482, when he went to Milan, he offered his services to the Sforza family in a long letter, most of which is devoted to describing his abilities as an engineer, skilled in the arts of war. He ends by saying:

In time of peace I believe I can give you as complete satisfaction as anyone else in architecture pertaining to the construction of buildings both public and private, and in conducting water from one place to another. Also, I can execute sculpture in marble, bronze or clay, likewise painting, in which my work will stand comparison with that of anyone else, whoever he may be . . .

We give a reproduction of one of his designs for a church (Plate 20). Regular polygons form a fundamental part of the design, and we shall consider it again when we discuss the notion of form in architecture, in Chapter 4. There is evidence that Leonardo did get as far as making a model of one of his proposed churches, and this was built, but he acquired a reputation for not completing works of art. It may be that during at least one period of his life, this was due to Leonardo's obsession with geometry.

Isabella d'Este attempted to commission a painting in 1501, using the good offices of Fra Pietro da Novellara, who was preaching in Florence then, but the friar reported that Leonardo was more devoted than ever to mathematics, which had so distracted him from painting, that 'he could not suffer the brush'.

Let us now examine some of Leonardo's excursions into geometry. Of course he was a superb practical geometer when he wished to be. But some of his geometrical diagrams are rather poor, even if we make allowances for the age of the manuscript, and he does seem to have become obsessed with matters which we no longer regard as important. Some sheets of manuscripts are covered with scores of rough diagrams appertaining to the areas of *lunes*, figures bounded by arcs of circles (see Plate 21). It was Hippocrates of Chios, in the fourth century B.C., who made the surprising discovery that certain lunes have an area that is calculable, that is, that squares having the same area can be constructed. This problem aroused great interest, since it was thought that if the area bounded by *two circular arcs* can be calculated, it must be possible to *square the circle*, that is to construct a square with area equal to that of a given circle, which is bounded by only *one* circular arc.

This hope was not fulfilled, but let us see what Hippocrates proved. The triangle ABC (Figure 29) has a right angle at C, and so the semicircle with AB as diameter passes through C. Semicircles are also described on AC and on BC as diameters. Hippo-

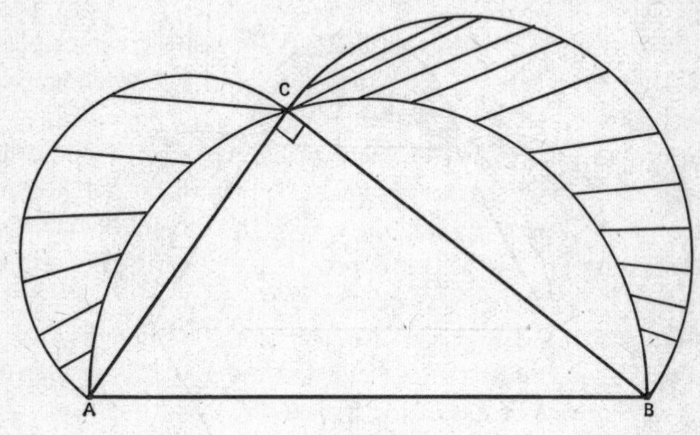

Figure 29

crates showed that the sum of the areas of the two lunes, shaded in the diagram, is equal to the area of the triangle ABC.

By Pythagoras, $AB^2 = AC^2 + BC^2$, and it was known that all semicircles are similar to each other, so that *their areas are proportional to the square of their diameters.* Hence

Area of semicircle on AB *as diameter*
= *area of semicircle on* AC *as diameter* +
area of semicircle on BC *as diameter.*

If we call the respective areas of the lunes L and M, the areas of the segments of circles shown x and y, and the area of triangle ABC is called Δ,

$$\Delta + x + y = x + L + y + M,$$

so that

$$L + M = \Delta.$$

This is the theorem which fascinated Leonardo, and, of course, it is possible to extend it, with semicircles drawn on the sides of squares as diameters, and so on, and this is what Leonardo did. Hippocrates has a theorem for lunes connected with a regular hexagon inscribed in a circle (Figure 30), that the sum of the areas

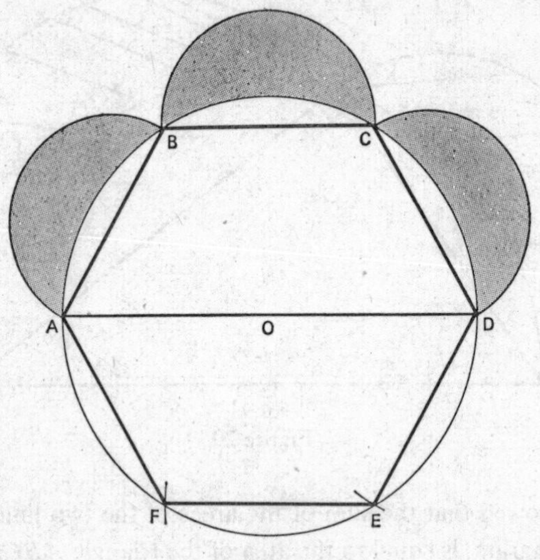

Figure 30

of the three lunes shown, plus that of one small semicircle is equal to half of the area of the original hexagon, and we leave the proof of this as an Exercise (p. 100).

The construction of an angle of 15° involves a fair number of circles, but Leonardo shows how to construct this angle with a rusty pair of compasses, one with a fixed aperture. The construction is based on the theorem that the angle subtended by an arc of a circle at the centre is twice the angle subtended at any point of the circumference. We show this theorem and Leonardo's construction in Figure 31. Draw a circle, centre *D*, choose any point *A* on this circle, and mark off, with the compasses, points *B* and *C* on the circle on either side of *A*. Then *ABD* is an equilateral triangle, and the line BC is the perpendicular bisector of AD. The circle centre B goes through A and D, and if it cuts the line BC in E, the angle ADE is the required angle of 15°. The proof is simple. The angle DBA is 60°, and BE bisects it, so that the angle ABE is 30°. The circle centre B passes through D, E and A, and

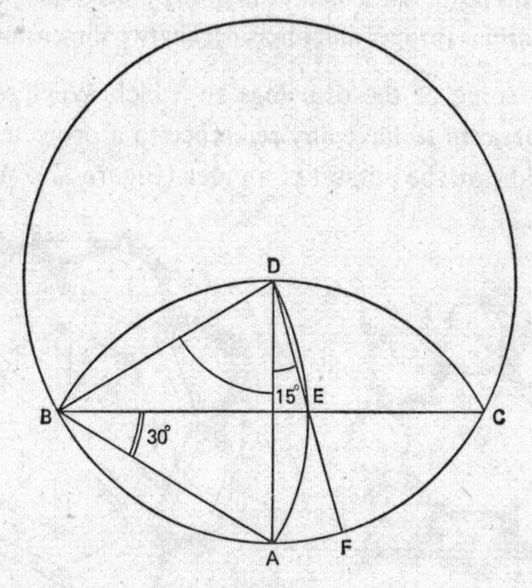

Figure 31

the arc AE subtends half the angle at D which it subtends at the centre of the circle, B. Hence the angle ADE is 15°.

Although Leonardo must have known the Elements of Euclid, where, as we saw (p. 71), the construction of a regular pentagon is given, he makes a number of attempts to construct a regular pentagon, and labels some of them *falso*. It is difficult to judge a man's capabilities from various doodles, but the opinion of some historians of mathematics has been very definite. They feel that Leonardo made no contributions to geometry worth recording. It is fairly clear that Leonardo was not very interested in theoretical constructions.

But now we quote Hermann Weyl. In his well-known book entitled *Symmetry* (Princeton University Press, 1952), p. 66, Weyl says:

> Leonardo da Vinci engaged in systematically determining the possible symmetries of a central building and how to attach chapels and niches without destroying the symmetry of the nucleus. In abstract modern terminology, his result is essentially our above table of the possible finite groups of rotations (proper and improper) in two dimensions.

We show some of the drawings to which Weyl refers, a set which do not seem to have any reference to a projected building, but which indicate the pursuit of an idea (Figure 32). We shall not

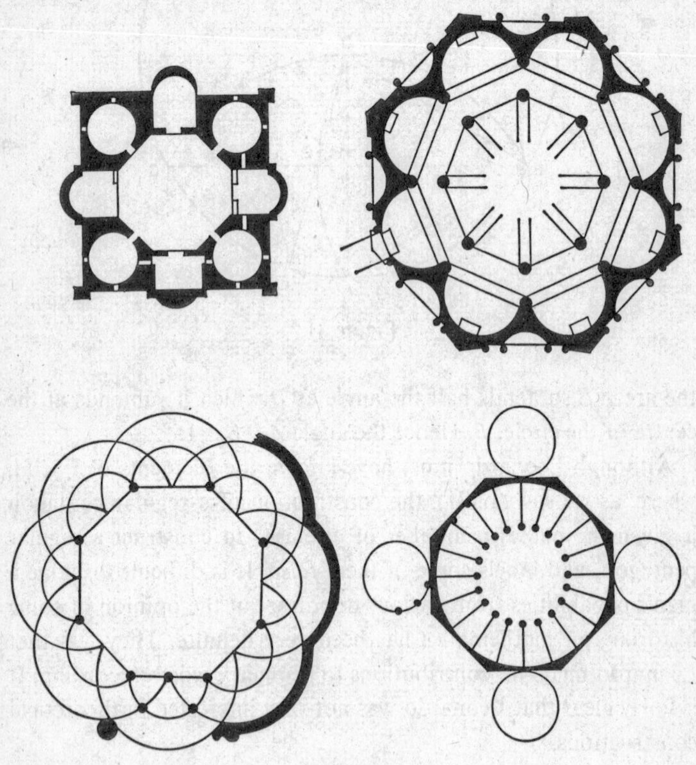

Figure 32

discuss the concept of *group* until Chapter 9. It is sufficient to remark here that although Leonardo seems to have restricted himself to a regular octagon (8-gon) inscribed in a circle, and Weyl is referring to a regular *n*-gon inscribed in a circle, Leonardo's work shows that he had a complete grasp of all the possibilities and this alone would make him important in the history of geometry.

Again, Leonardo gives an interesting way of tracing an ellipse. It is known that if we make a triangular cut-out ABC, draw two lines l and m on a sheet of paper, lay the triangle ABC down so that the vertex A lies on l, while the vertex B lies on m, and mark the position of the vertex C on the paper, then the various positions of the vertex C will trace an ellipse (Figure 33).

Leonardo was interested in another aspect of this construction. He imagined a fixed stylus at the point C, and moved the plane beneath, on which the lines l and m are marked, so that l always passes through the fixed point A, and m always passes through the fixed point B. The stylus at C will then trace out part of an ellipse. This method has applications to linkages, camshafts, and so on.

Finally, Leonardo is associated with the discovery of the centre of mass of a tetrahedron. To explain this concept, we begin with a triangle ABC which we imagine cut out of uniform material, sheet metal, say. Is there a point G associated with this triangle which is such that if we drive a nail through G into a wall, the triangle will hang freely in any position? Such a point is called *the centre of mass* of the triangle, and it can be found by joining the point A to the centre of BC, the point B to the centre of CA, and the point C to the centre of AB. These three lines, called *medians* of the triangle ABC, all pass through the same point, and this is G. It is one third of the way up the line joining the midpoint of any side to the opposite vertex (Figure 34).

This theorem was known to Archimedes, and Leonardo is known to have expressed a great admiration for the works of Archimedes, which were available at that time. The original

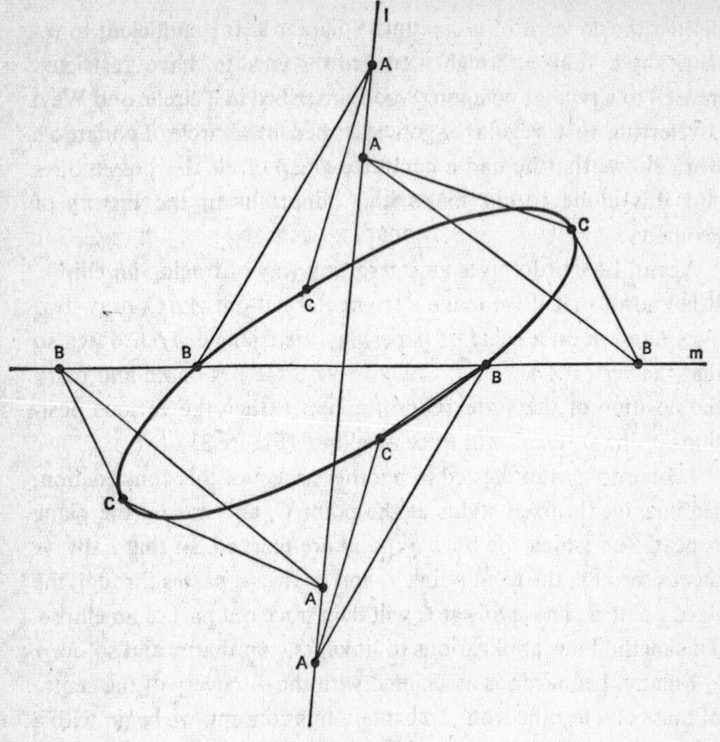

Figure 33

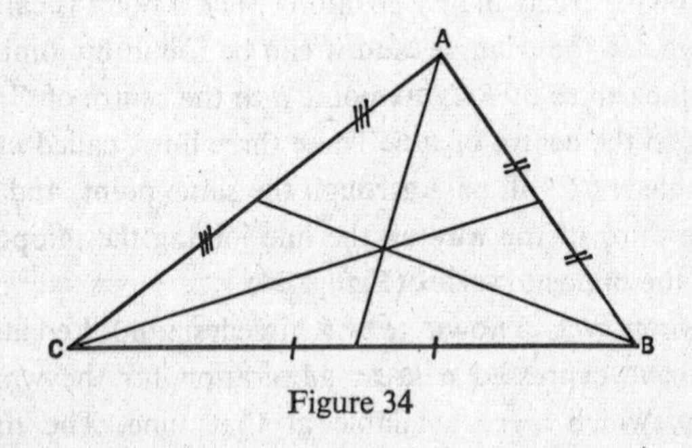

Figure 34

contribution Leonardo is thought to have made is in determining the centre of mass of a uniform tetrahedron, that is of a uniform *solid* tetrahedron. Leonardo states, correctly, that this is to be found *one quarter* of the way up from the base along a line joining the centre of mass of the triangular base to the opposite vertex. The implied theorem is that all these lines pass through the same point (Figure 35). Leonardo does not give a proof, and

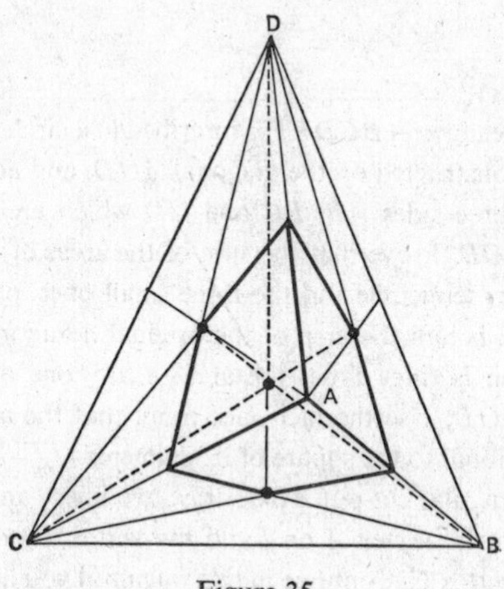

Figure 35

again we do not know whether he was merely noting a result he had come across.

We began this chapter by admitting that the Florentine Leonardo is one of history's enigmas. Fairly recent historical research has shown that some of the ideas in his Notebooks which seem to be so startling and original are merely transcriptions from books which appeared in Leonardo's time. If Leonardo was merely making notes, it is not his fault that posterity has rushed in to hail him as an amazing inventor. The air was dense with ideas and inventions during the Renaissance, and Leonardo had an enormous curiosity.

On the other hand, the marvellous drawings in the Notebooks are certainly his own, and there exist a fair number of paintings which are indubitably from his brush, and these are also magnificent. Leonardo was much appreciated in his own time, as we have seen. He did not know enough geometry to be able to cope with rather difficult problems, but as we shall see in Chapter 5, p. 136, when he was unable to solve a theoretical problem in optics, he constructed a mechanism which gives a practical solution.

Exercises

1. A regular hexagon $ABCDEF$ is inscribed in a circle, centre O, and a semicircle is constructed on the diagonal AOD, and additional semicircles on the three sides AB, BC and CD which are chords of the semicircle on AOD. Prove that the sum of the areas of the three lunes between the large semicircle and the three small ones, plus that of one small semicircle, is half the area of the original hexagon. (The area of half the hexagon is three times the area of any one of the triangles AOB, BOC, COD. Use the fact once more that the area of a semicircle is proportional to the square of its diameter.)

2. Make a triangular cut-out ABC, draw two lines l and m, and place the cut-out with the vertex A on l and the vertex B on m. Mark the position of the vertex C. Continue in this way until you have a sufficient number of points to indicate the path traced out by the vertex C. Is it an ellipse?

3. As a special case of Exercise 2 above, draw l perpendicular to m, let C be a fixed point on a line segment AB, and trace the position of C by pricking through the paper as A moves on l, and B moves on m. You should obtain an ellipse with l and m as principal axes.

4. As a variation on the above, using the same triangular cut-out ABC, draw two circles c and d, and position triangle ABC so that BC touches the circle c and CA touches the circle d. Trace the third side AB so that it comes through onto the paper. Continuing thus, the various lines AB should touch another circle.

5. Let O be the centre of a segment AB two units in length, and draw a semicircle on AB as diameter. Let the perpendicular through O to AB

meet the semicircle in C, and complete the squares $OCKB$ and $OCIA$ so that $ABKI$ is a rectangle, and IK touches the semicircle at C. Draw the circle with centre at the midpoint of OC to pass through C, and therefore O, and semicircles with centres at the midpoints of OB and OA respectively to cut this circle in G and F respectively. Draw the quadrant of a circle, centre K to pass through C and B and to lie inside the rectangle $ABKI$, and similarly the quadrant of a circle centre I which passes through C and A. Now complete the figure by drawing a quadrant of a circle, centre at the midpoint of CI, to pass inside the rectangle through C and F, and a quadrant of a circle, centre at the midpoint of IA to pass inside the rectangle through F and A, and draw similar quadrants on the other side of the axis of symmetry CO.

Leonardo investigated the areas of the larger and smaller lunes in this figure, which is in itself an attractive design. See if you can produce some more.

·4·

FORM IN ARCHITECTURE*

In previous chapters we have already discussed the appearance of things, in architecture and in painting, and we shall now discuss the theory of proportion in architecture. Can purely formal relationships in the design of buildings give pleasure? There are some who would deny this possibility. It has even been asserted that the whole problem of proportion in architecture can be reduced to structural mechanics, to the implied necessity of the structure, and moreover that engineering skill can produce a unique, perfect and automatically pleasing solution to every problem of design. In other words, structure plus utility constitute the *fitness* which played an important part in debates on aesthetics and the theory of proportion in the eighteenth century. An attempt was often made by critics and philosophers to reduce the theory of proportion to a simple theory of fitness.

We may think this theory is as capable of manipulation by interested parties as a theory of *significant form*, which dominated aesthetic theory some years ago. If we disregard theories, for the moment, it is clear that custom and convention play an enormous part in our appreciation of buildings. If the shapes which are more pleasing to the eye can be selected (and this is where the difficulty lies), architectural proportion can become a straightforward matter of using these shapes as often as possible.

*In this chapter I have made full use, with permission, of P. H. Scholfield's excellent book: *The Theory of Proportion in Architecture* (Cambridge University Press, 1958).

In the Renaissance it was commonly believed that the most beautiful rectangles were those whose sides had the simple numerical relations of musical consonance. These notions were ascribed, mistakenly, to Vitruvius. The Greeks, although a highly literate people, left no surviving records which describe the system of proportions they used in their architecture. At a later date, the only surviving literary evidence appears in Vitruvius, and, as we have remarked, translators have often been aware that the meaning of a number of words used by Vitruvius has been lost. There were no commentaries during the medieval period, and so the study of Vitruvius had to make a completely fresh start in the Renaissance. The humanists of the early Renaissance were neo-Platonists, and regarded Plato's work as the flower of classical culture. Few of them made much sense of Vitruvius' writings on proportion in architecture. In fact it was the mathematician Cardano, in the sixteenth century, who attributed to Vitruvius a theory of proportion based on music. Although this cannot be found in Vitruvius, there is a discussion in Plato's *Timaeus*. Vitruvius' interest in music was confined, as we saw (p. 17), to the correct tuning of the stretched chords of catapults, and to the construction of resonators in theatres, to help audibility.

The Renaissance architect Alberti, in his *Ten Books on Architecture*, wrote:

> . . . and indeed I am every day more and more convinced of the truth of the Pythagoras saying, that Nature is sure to act consistently, and with a constant analogy in all her operations: from whence I conclude that the same numbers by means of which the agreement of sounds affects our ears with delight, are the very same which please our eyes and mind.

Alberti used these musical proportions to relate the three dimensions, height, length and breadth, but not in adding or subtracting dimensions. The musical analogy stressed the pleasure given to the eye on observing ratios which are expressible as *ratios of integers*, and Cardano maintained that such ratios are pleasing because they are intelligible. But when these matters were being discussed in the nineteenth century, it was pointed out that an octave is the most perfect of all concords in music, but the ratio of two to one (that of the frequencies of the notes involved) is hardly an agreeable one in a building.

Earlier than this Hogarth, in his *Analysis of Beauty* (1753), talked about writers who 'have not only puzzled mankind with a heap of minute and unnecessary divisions, but also with a *strange notion* that these divisions are governed by the laws of music'. And Helmholtz, in his *Sensations of Tone*, published in 1862, observed that the consonance of two notes of a chord can be explained quite simply by the absence of unpleasant beats between the partials of the notes which are reproduced by certain membranes of the human ear. It is thus primarily a physiological phenomenon, depending on the structure of the ear, and not a psychological phenomenon depending on the recognition of simple ratios by the mind itself. In other words, an alteration of the membranes would produce a change of aesthetic attitude! Unfortunately, we are far less certain nowadays than Helmholtz was as to what we mean when we talk of *a purely psychological phenomenon.* We shall not venture any further into these disputed realms, but merely wish to stress the great importance of *theory of some kind* in architectural design.

Palladio (1518–1580), a century after Alberti, when the first impetus of the Renaissance was expended, was rather cautious in any theoretical explanations of what he did. The counter-reformation stressed the importance of authority, and this meant Vitruvius, and of deductions made from archaeological measurements of the physical remains of antiquity. Where Palladio does depart from Vitruvius, he relies on the measurements of classical buildings, but he avoids the musical analogy, and Alberti's other ideas, except for the use of arithmetic, geometric and harmonic means when he determines the height of vaulted rooms.

We briefly interpose the definitions of these means. A set of numbers $a_1, a_2, a_3, \ldots$ is said to be in *arithmetic progression* if the difference between any number and its successor is constant, so that

$$a_2-a_1 = a_3-a_2 = a_4-a_3 = \ldots$$

For example, the set 1, 3, 5, 7, 9, ... is in arithmetic progression. For such a set $a_1+a_3 = 2a_2$, $a_2+a_4 = 2a_3$, ... and so on, and every number in the set is equal to *one half the sum* of the numbers on either side of it. Any number in the set is said to be *the arithmetic mean* of the numbers on either side of it, and so we have the rule for forming the arithmetic mean of two given numbers a and b,

$$\text{arithmetic mean} = (a+b)/2.$$

If a set of numbers $a_1, a_2, a_3, a_4, \ldots$ is such that the ratio of any number of the set to its predecessor is constant,

$$a_2/a_1 = a_3/a_2 = a_4/a_3 = \ldots$$

then the set is said to be in *geometric progression*. Here $a_1a_3 = (a_2)^2$, $a_2a_4 = (a_3)^2$, ... and so every number in the set is equal to *the square root* of the numbers on either side of it, and *the geometric mean* of two given numbers a and b is given by the formula:

$$(\text{geometric mean})^2 = ab.$$

Finally, a set of numbers $a_1, a_2, a_3, a_4, \ldots$ is said to be in *harmonic progression* if the set of *inverses*

$$1/a_1, 1/a_2, 1/a_3, 1/a_4, \ldots$$

is in *arithmetic progression*. Again, any number of the set is called the *harmonic mean* of the numbers on either side of it, so that to find the harmonic mean of two numbers a and b, we first find the arithmetic mean $(1/a+1/b)/2$ of their reciprocals, and the reciprocal of this, so

$$\text{harmonic mean} = 2ab/(a+b).$$

The observation that the pitch of a stretched string depends on the inverse of its length goes all the way back to Pythagoras. If a stretched string is plucked, and then stopped halfway, and plucked again, the note given out the second time is the octave of the unstopped note. If the string is stopped so that only a third of it vibrates, the note given out has a frequency equal to three times the fundamental frequency, and the notes of a scale can be formed by stopping the string at points which are rational multiples of the original length. Hence the importance of the *harmonic* series, the inverses of which are in arithmetical progression. The simplest harmonic series is, of course, 1, 1/2, 1/3, 1/4, . . .

As we remarked above, Palladio did use arithmetic, geometric and harmonic means for determining the height of vaulted rooms. Architectural ideas suitable for Italy were sometimes taken over in colder countries, with disastrous consequences as far as comfort is concerned, producing rooms with high ceilings which cannot be adequately heated. Palladio used a simpler rule for rooms with flat ceilings. He insisted that they should be as high as they are broad.

The High Renaissance arrived so late in England that the musical theory of proportion survived into the eighteenth century without requiring a revival. It was then laughed out of court just at the time that it was being revived in France and Italy, and, as we saw, Hogarth helped to destroy any credence it may have gained. But it certainly had its day, and Inigo Jones (1573–1652) used it in designing the Banqueting House in Whitehall, which is a double cube, and the Queen's House at Greenwich, which has simple integral proportions for its rooms.

Two squares put together form a double square, and if two double squares are put together we obtain a square, which repeats the pattern of the original square. This simple additive property of the square was used effectively in Renaissance architecture (Plate 22).

Dürer tried both the harmonic and the arithmetic scales of

proportion, the harmonic scale to take one half, one third, and so on of a unit, and the arithmetic scale merely to add units together. The interesting thing about Dürer's extensive researches on human proportions is that *the one example* of the repetition of ratios which he says he found involved the measurements *neck to hip*, *hip to knee* and *knee to ankle*. Dürer believed that he found these to be in geometric progression, so that for most human beings:

$$(\text{neck to hip})(\text{knee to ankle}) = (\text{hip to knee})^2.$$

There were attempts, following the Renaissance, to prove that the architects of that glorious period had never used incommensurable proportions, that is proportions which are *not* expressible as the ratio of two integers. But even Vitruvius advocates the use of rectangles in which one side is $\sqrt{2}$ times the other, and it is no doubt for this reason that in 1570 Palladio includes the $\sqrt{2}$ rectangle in the list of seven shapes which he recommends for the plans of rooms.

Expanding star pentagons do not seem to have been used, and we have remarked on the almost universal absence of pentagons in architecture (p. 68). The connection of the pentagram with magic may have had something to do with it. But there is no doubt at all about the use of the eight-pointed star (Plate 20 and Figure 44). This design by Leonardo is just one of the many in his Notebooks and nearly all are based on expanding eight-pointed stars. The division of a circle into eight equal parts produces an angle of 45°, and an isosceles right-angled triangle with unit sides has angles of 45° and hypotenuse equal to $\sqrt{2}$, so eight-pointed stars certainly introduce irrational numbers.

The golden ratio, which we discussed in connection with the Euclidean construction of a regular pentagon (p. 71) was known to Renaissance architects, but no effective use was made of it as an instrument of proportion. There was, however, enormous interest in it, and Piero della Francesca and Luca Pacioli, in their studies of the five regular Platonic solids, showed that they had learned all that Euclid could teach them. We know that Pacioli referred to the golden ratio as the *divina proportione*. The term *golden ratio* is thought to have originated in Germany in the first half of the nineteenth century. It is this ratio $\mu = (1+\sqrt{5})/2$ which we shall be talking about for a time.

Is a rectangle with sides in the golden ratio more beautiful than one whose sides are in the ratio 2:1, say, or 3:2, or 5:7? Experiments have been conducted to decide this question, with findings which are not altogether convincing, but with a slight preference, perhaps, for the golden ratio. But then can a rectangle *all by itself* be outstandingly beautiful or depressingly ugly?

The secret of proportions seems to lie not in single shapes, but in the relationship between them. We have hinted already that theories of aesthetics, which naturally involve what one sees, bring out the fiercest passions in apparently mild people. In any architecture library, books attacking the golden ratio as an architectural tool can be read, and some are very sensible.

One finds the very pertinent remark that everything in architecture, as in painting, depends on the position of the observer, and that if some proportions in a building appear to be in the golden ratio from one position, they will appear in a different ratio from another. We amplify this statement, introducing some geometry.

Suppose that the front of the building is a rectangle (Figure 36), divided by a horizontal line in the golden ratio, so that if we take a plane section orthogonal to the façade $CB/BA = \mu$, the golden ratio. We also suppose that an eye moves along the horizontal through A, and that the observer feels an aesthetic thrill at the point F, which means that, since apparent size depends on the angle subtended at the eye,

$$(\text{angle } CFB)/(\text{angle } BFA) = \mu.$$

Now let the circle through C, F and B intersect the horizontal through A again at E. Then angle CEB = angle CFB, these be-

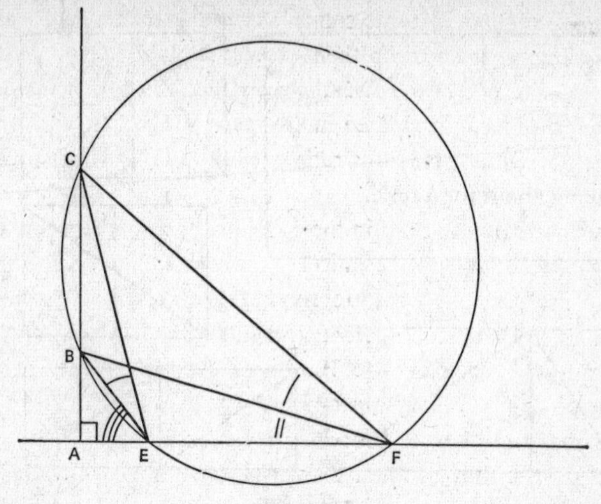

Figure 36

ing angles in the same segment of a circle, but angle BEA, being the exterior angle of triangle BFE, is greater than the angle BFA. Hence (angle CEB)/(angle BEA) $<$ (angle CFB)/(angle BFA), so that if the division by a horizontal line looks like the golden ratio in one position, it will not in other positions.

Hermann Weyl, in his book *Symmetry*, to which we have already referred, merely mentions 'the *aurea sectio*, which has played such a role in attempts to reduce beauty of proportion to a mathematical formula'.

But something can be said for the aesthetic appeal of certain ratios, the golden ratio being one, and we now consider the treatment given by P. H. Scholfield in his book, *The Theory of Proportion in Architecture* (Cambridge University Press, 1958).

In Figure 37a we see a rectangle divided into two rectangles by a line parallel to two sides. We compare the shapes of the rectangle with the shapes of the two rectangles into which it is divided. In general the three shapes are different. But there are ways of eliminating one of the shapes. In Figure 37b one of the smaller

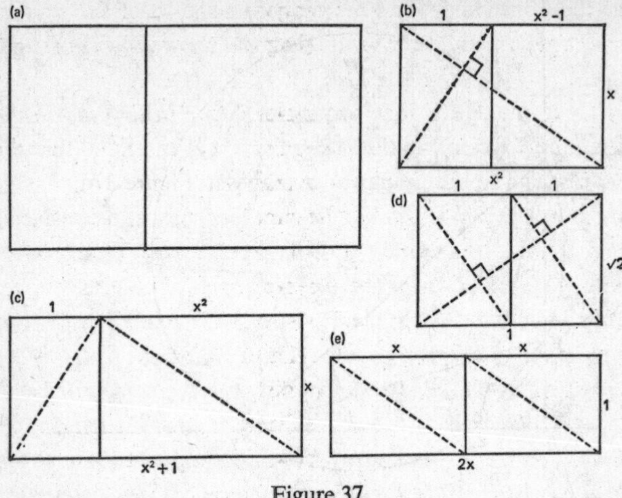

Figure 37

rectangles is similar in shape to the original one. The sides of the rectangle are x^2 and x, so that the ratio is x, and the division on the longer side is at a distance 1 from an edge. The ratio for the smaller rectangle is therefore also $x:1 = x$, so that this has the same shape as the original rectangle, whatever x. We notice that the two diagonals shown are perpendicular.

In Figure 37c, the original rectangle has sides x^2+1 and x, and the division is again one unit from an edge. Now the two rectangles formed by the division are similar to each other, the ratio of the sides being $x:1 = x$, but they are not necessarily similar to the original rectangle. In Figure 37e, the sides of the rectangle are $2x$ and 1, and the division is at the midpoint, so that the two rectangles obtained are identical.

We can now eliminate two of the three shapes present in each figure if we consider Figure 37b again and write

$$1/x = (x^2-1)/x,$$

which gives us the equation $x^2 = 2$. If we now take $x = \sqrt{2}$, Figure 37b becomes Figure 37d, in which the three rectangles all have the same shape, and the diagonals shown of the smaller rectangles are perpendicular to the diagonal shown of the original rectangle.

If we try to make the three rectangles shown in Figure 37e the same shape, we obtain Figure 37d again. It will be agreed that the eye does respond to the similarity of shapes in Figure 37d.

We continue this investigation of simple shapes, and consider a rectangle divided by parallels to both sides, as in Figure 38a. If we count, we find that there are *nine* rectangles of different shapes in the figure. In Figure 38b, one division is a central division, and this reduces the number of distinct shapes from nine to six. If the problem of reducing the number of distinct shapes still further is studied, there are three cases where the number of distinct shapes is reduced to *three*, and these are shown in Figures 38 c, d and e, where $\mu = (1+\sqrt{5})/2$, the golden ratio. In checking the similarity of shapes, it is useful to use the property $1+\mu = \mu^2$ of the golden ratio.

There is also a case, shown in Figure 38f, where there are no horizontal or vertical axes of symmetry, just a diagonal one. The original shape is a square. There are three squares in the figure, and six rectangles, two of these having the ratio $1+\mu = \mu^2$ for their sides, and the other four having the ratio μ.

The golden ratio does appear to have some attractive properties, and we investigate its mathematical properties further.

If we consider the geometric progression

$$1, \mu, \mu^2, \mu^3, \mu^4, \mu^5, \ldots, \mu^n, \ldots$$

then since

$$\mu^3 = \mu+\mu^2 = \mu+(1+\mu) = 2\mu+1,$$
$$\mu^4 = 2\mu^2+\mu = 3\mu+2,$$
$$\mu^5 = 3\mu^2+2\mu = 5\mu+3,$$

and so on, we finally see that from $\mu^n = \mu^{n-1}+\mu^{n-2}$, the *co-efficients* of μ which we obtain when we express powers of μ in terms of μ are the series of integers

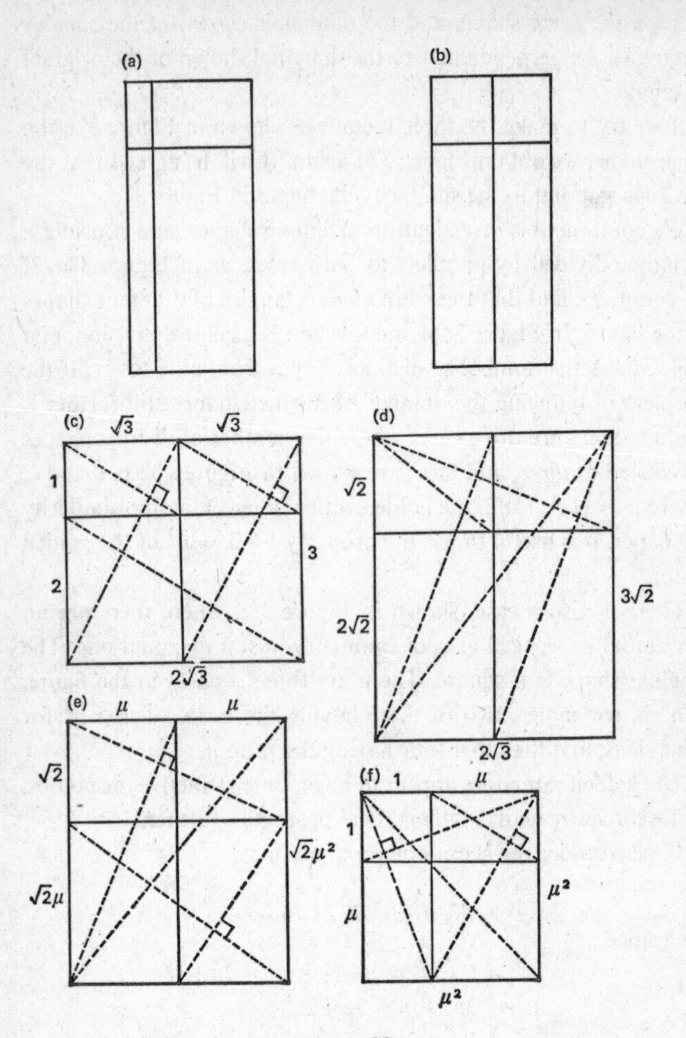

(a)
(b)
(c)
$\sqrt{3}$
$\sqrt{3}$
1
2
3
$2\sqrt{3}$
(d)
$\sqrt{2}$
$2\sqrt{2}$
$3\sqrt{2}$
$2\sqrt{3}$
(e)
μ
μ
$\sqrt{2}$
$\sqrt{2}\mu^2$
$\sqrt{2}\mu$
(f)
1
μ
1
μ^2
μ
μ^2

Figure 38

$$1, 1, 2, 3, 5, 8, 13, \ldots$$

in which if u_n is the nth term of the series,

$$u_n = u_{n-1} + u_{n-2}.$$

Thus, beginning with the third term, $2 = 1+1$, and then $3 = 2+1$, $5 = 3+2$, $8 = 5+3$, $13 = 8+5$, and the next term of the series is $21 = 13+8$, and so on.

This set of numbers has a long history, and is called *the Fibonacci series*. Leonardo of Pisa came across it in 1202, in connection with the breeding of rabbits! His nickname was Fibonnaci, son of good nature, and this has stuck to the series.

He assumed that rabbits live forever, and that every month each pair begets a new pair, which become productive at the age of two months. In the first month the experiment begins with a newborn pair of rabbits, so that we write down the number 1. In the second month there is still just one pair, so we write down 1, again. In the third month, a pair is born, so we write down the number 2. In the fourth month we have 3 pairs, in the fifth month 5 pairs, and so on, and it is easy to see how the following relation arises:

$$u_n = u_{n-1} + u_{n-2}.$$

Curiously enough this relation was not actually stated by Leonardo of Pisa, but it was noted by Kepler, four centuries later.

The golden ratio, as we saw, arises in connection with the regular pentagon, and therefore the Fibonacci series also plays a role in anything relating to regular pentagons or star pentagons. In Figure 39 we show a few of the various ratios which involve the golden ratio, and which excited Pacioli so much, and numerous artists and architects since.

It is the *additive properties* of the golden ratio which are important in design. Each time a new length u_n of the design is drawn equal to μu_{n-1}, where the length u_{n-1} is already drawn, there is

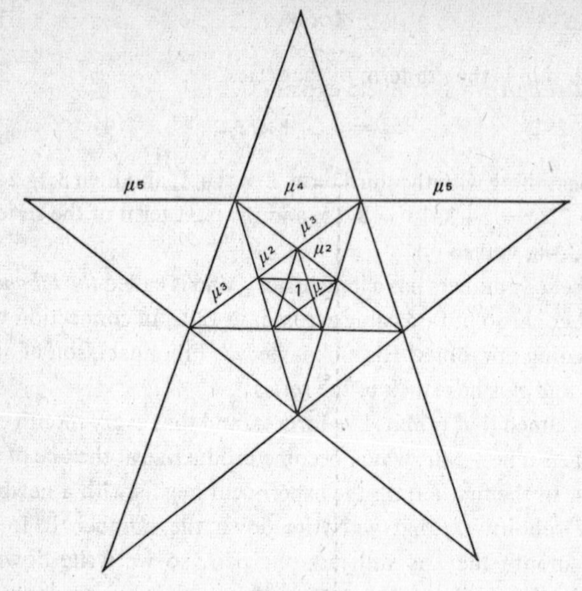

Figure 39

the satisfaction of knowing that $u_n = u_{n-1}+u_{n-2}$, so that parts of the design can be fitted together.

The enthusiasm shown by disciples of the golden ratio is illustrated in Plate 23, which is taken from *The Curves of Life*, by Sir Theodore Cook, which was published in London in 1914. The demure lady is, of course, a Botticelli Venus, and Sir Theodore has chosen certain aspects of her form for demonstrating that the distances between them are in the golden ratio. One might argue with Sir Theodore about certain points in the figure being preferred to others. Vitruvius, it will be remembered (p. 24), chose the navel as one key point, and Sir Theodore follows precedent to that extent. Sir Theodore, and many others, use a different symbol to denote the golden ratio where we use μ.

Sir Theodore also considers a Turner seascape, and gives *A Universal Golden Ratio Scale*, which P. H. Scholfield quite rightly terms a forerunner of the Modulor, the scale used by Le Corbusier. A little simple geometry is involved, which neither Sir Theodore nor Mr Scholfield explain:

In Figure 40 let A, A', A'', . . . be points on a line, and suppose

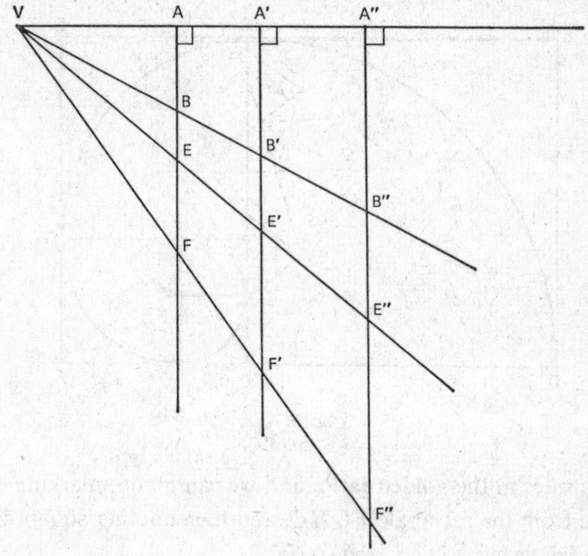

Figure 40

that the lines AB, $A'B'$, $A''B''$, . . . are all perpendicular to the line AA', where the lines $BB'B''$, $EE'E''$, $FF'F''$, . . . all pass through the same point V on the line AA'. Then it is a theorem of similar triangles that the ratios $AB{:}AE{:}AF{:}\ldots$, $A'B'{:}A'E'{:}A'F'{:}\ldots$, $A''B''{:}A''E''{:}A''F''{:}\ldots$ are all equal.

It is also a special case of this theorem that since the lines AB, $A'B'$, $A''B''$, . . . are all parallel, the ratios $AA'{:}AA''{:}\ldots$, $BB'{:}BB''{:}\ldots$, $EE'{:}EE''{:}\ldots$ are all equal. Hence we have a method for constructing any number of scales in the golden ratio, with a unit length suitable to the construction or verification we are undertaking. In Exercise 3 at the end of this chapter the reader is encouraged to make his own Universal Scale.

In Figure 41 we show a rectangle in golden ratio with a square *ABCD* cut off, which leaves a rectangle *BEFC* with sides also in the golden ratio. If we now take this rectangle *BEFC*, and mark off a square *BEGH* from it, the remaining rectangle *FGHC* also

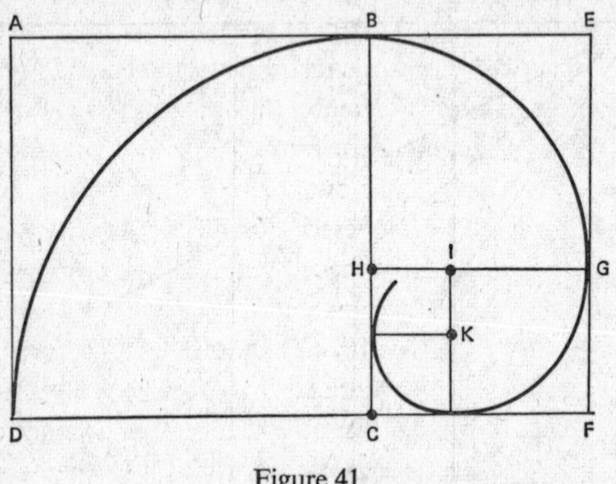

Figure 41

has its sides in the golden ratio, and we can go on, marking off a square from the rectangle *FGHC*, and then another square from the rectangle left behind, and so on.

If we use these successive squares to draw quadrants of circles, as we did when demonstrating Dürer's approximate construction of a spiral (Figure 41), we obtain a very attractive curve. The successive centres of the quadrants are *C*, *H*, *I*, *K*, . . . In Chapter 8 we shall consider a true spiral, *the equiangular spiral*, which is related to the golden ratio.

Having shown some enthusiasm for the golden ratio, we must now stress that for every enthusiast there has also been a severe critic. Ruskin, in *Modern Painters*, said: '. . . the determination of right or wrong proportions is as much a matter of feeling and experience as the appreciation of good musical composition.' Three years later, in 1849, in *The Seven Lamps of Architecture*, Ruskin disparaged the use of proportions in architecture, comparing it to 'the disposition of dishes at a dinner-table, of ornaments on a dress'. Ruskin was a very influential critic, and the collapse of the Renaissance theory of architecture, and the various theories of proportion which formed part of it left the architects of nineteenth-century England with the largest building programme the world had yet seen, and with no coherent principles to guide them.

The Gothic revival was a natural development. We need only mention the House of Commons at Westminster as an example of a move forward in time from patterns based on ancient Greek buildings, with Doric, Ionic and Corinthian columns. But a return to the Gothic hardly offered a set of rules for building, since very little was known about medieval theories of proportion.

It is obvious that since translations of Euclid from the Arabic were available from the twelfth century onwards, geometrical constructions were used, but there is little information on how they were used. One of the few bits of evidence is an account of a dispute over the design of Milan cathedral in 1392, when a majority of the architects were in favour of carrying out a triangular rather than a square design.

The one clear account of Gothic methods which has survived gives three rules for designing churches. The first fixed the overall length and breadth of the church by means of the construction given in Euclid's Proposition 1, the erection of an equilateral triangle on a given base (Figure 42). The length is represented by CF and the breadth by AB, and the shape of the area bounded by the two circles was often referred to as a *vesica piscis* (fish bladder). But this construction, which also gives a method for constructing a perpendicular to a given line, may merely have been used to set out the main axes of the church, since not all Gothic churches have the relative dimensions of the vesica piscis. The second extant Gothic rule provided for the subdivision of the plan of the church into equal bays, and the third determined the height of various parts by means of the construction of equilateral triangles.

Of course the construction of rose windows, as at Chartres,

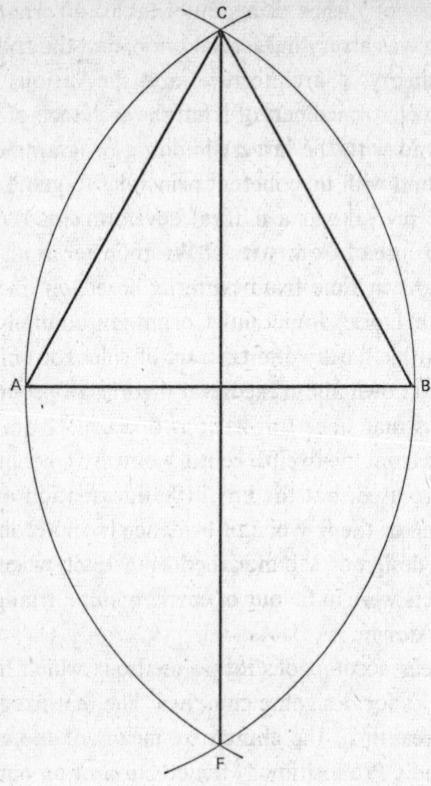

Figure 42

involved the geometry of regular and star polygons, and some sketchbooks remain to throw light on the subject, but while an interest in geometry is fairly obvious, there is little indication of its use. Research into the dimensions of ancient ruins usually produced the kind of results the investigator hoped to obtain, and similarly much of the research into Gothic architecture, with the fitting together of circles and triangles to prove a contention, was controversial. And of course, even Vitruvius had recommended subtle optical corrections, so if the Gothic architects had altered their designs as they went along, and the cathedrals of the Middle Ages took many years to build, almost any theory could be tailored to fit the facts.

The difficulty in discovering form in architecture can be illustrated by the history of St Paul's Cathedral, in London. This was built after the Great Fire had made restoration of the original Gothic structure impossible. The architect, Christopher Wren, was Professor of Astronomy in Oxford in 1657, and although he spent some time in Paris, he never managed to visit Italy. While in Paris he saw parts of the Louvre being built, and met the great Bernini, who was in charge of the building. Wren remarks that all that Bernini showed him with respect to the Louvre plan was a set of five little designs on paper, which Wren noted he would dearly have loved to copy.

The original designs for St Paul's made by Wren were rejected, and there are no designs of the present structure. It seems that Wren first intended a plan modelled on the Greek cross, but the Commissioners felt that 'a long body with aisles' would be thought 'impertinent, our religion not using processions'. Besides also asking for 'a dome conspicuous above the houses' (a phrase quoted without effect when skyscrapers began to be erected in London), the Commissioners did not want a plan which 'deviated too much from the old Gothick form of Catholic Churches which they (the people) had been used to see and admire in this country'.

Wren's dome is evidently Italian in style, but the construction is quite different from that of St Peter's, in Rome. Since Wren also rebuilt many churches in the City of London, he had ample opportunity to evolve a theory of form in architecture, if he had wished to do so, but the fact that he literally learned about architecture while working on the scaffolding with the masons may have kept him too busy to evolve a theory.

To return to the Victorian period, the situation in England in the second half of the nineteenth century was thus: architects were disappointed with arithmetic, and distrustful of geometry, and some therefore broke away and created their own styles. Victorian architecture is now appreciated in England, but of course there is not too much of it left to admire.

Have we discovered any basic principle of satisfactory design in the course of our investigations? Whatever the principle, it will be attacked, but there have been advocates for *the repetition* of similar figures. We have already referred to those who are ready

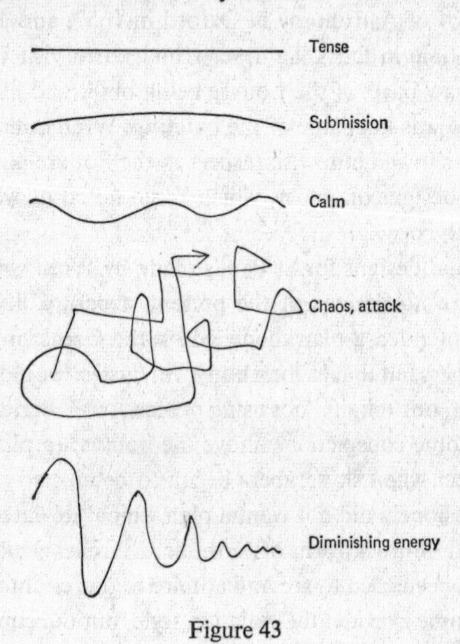

Figure 43

to attack the golden ratio if used as a static principle. Some point out that the perception of a building is a dynamic process and architectural *rhythm* is the important factor. We give an illustration of the effect produced on observers by a number of simple line drawings (Figure 43). Perhaps not everyone will interpret these drawings in the same way, but it is certain that a common feature of many buildings generally considered attractive is the repetition of similar geometrical figures.

We conclude this chapter with a rather brief mention of Le Corbusier, whose system conforms to that advocated by some architects, since it uses a scale, the Modulor, invented by Le Corbusier, which ensures the repetition of similar shapes. The Modulor consists of two scales, the *red* and the *blue* scale. The dimensions of the blue scale are double those of the red, and the divisions of each scale are based on the μ series, where μ is the golden ratio. Thus the Modulor is not only an instrument of architectural proportion, but a means of ensuring the repetition of similar shapes, as we saw in our discussion of the various shapes found in a rectangle with horizontal and vertical transverse lines (p. 109).

The unit length in any scale is of importance, and we saw that the Modulor is based on the human figure. Another module used by Le Corbusier is the figure of a man with his arm raised above his head. These modules were used quite successfully in the design of furniture, as well as of buildings, and anthropometric measurements are always to be commended in furniture design. So much so-called Danish furniture, armchairs, couches and so on, seems to be designed for Headless Wonders, since there is nowhere to lay one's head.

Le Corbusier was very successful, and he helped to design the United Nations' building in New York City. One wonders whether he ever read Vitruvius? The designers of the giant building in Manhattan ignored the fact that the afternoon sun, shining into countless office windows, would heat up the interiors in an unbearable fashion. Vitruvius, who was very concerned with human comfort, would never have made such a mistake.

Le Corbusier, who is not easy reading for those who wish to understand what they read, does raise an interesting point. Assuming that the Modulor is 'the key to the door of the miracle of numbers', if only in a very limited sphere, is that door merely one of hundreds, or did he, by sheer luck, open the one and only door beyond which there is a worthwhile discovery?

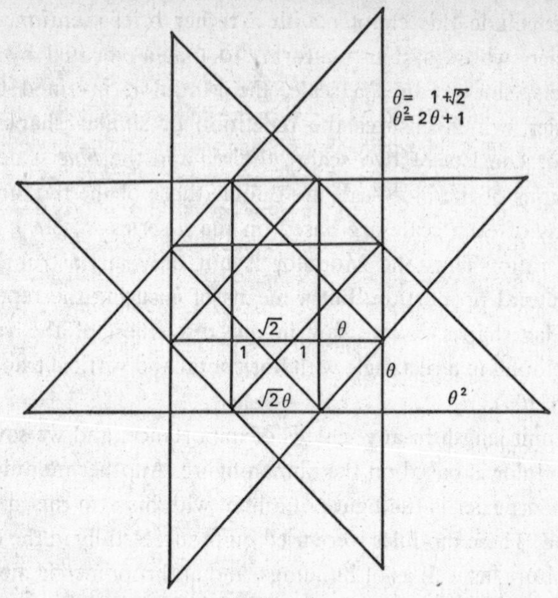

Figure 44

Our last drawing in this chapter (Figure 44) shows another possible Modulor, with a ratio $\theta = 1+\sqrt{2}$, as illustrated in a Da Vinci star-shaped octagon. For the ratio θ, we have $\theta^2 = 2\theta+1$, and if

$$u_n = \theta u_{n-1} = \theta^2 u_{n-2},$$

we obtain the relation $u_n = 2u_{n-1}+u_{n-2}$, which contrasts unfavourably with the relation $u_n = u_{n-1}+u_{n-2}$ between successive terms in the golden ratio. The simple additive property of the golden ratio is unique.

Exercises

1. ABE is a segment divided in the golden ratio at B, so that $AE = \mu AB$ (Figure 41). A square $ABCD$ is erected on AB, and a

square $AEGI$ on the same side of AE. The line BC produced meets IG at H, and DC produced meets EG at F. Show that the following four triples of lines each pass through a point: EH, AF, BG: EH, AG, IF: GA, BI, ED: and FI, GD, AH. Mark the lines in the figure which are perpendicular to each other. Do you find the figure an attractive one? Repeat the construction, with another ratio, say $k = 1{\cdot}5$, so that $AE = kAB$. You will not find the same concurrence of line triples.

2. Draw a regular pentagon with a side equal to the segment AB of Exercise 1 above, and extend the sides so that you obtain a star-shaped pentagon. Verify that the distance from A to the nearer vertex of the star which lies on AB is equal to the segment AE of the above Exercise. Verify that a side of the regular pentagon constructed from the star-shaped pentagon is also equal to AF, where F is the other star vertex on AB. Using the relation $1+\mu = \mu^2$, where μ is the golden ratio, see if you can obtain the relations between the sides of the pentagons and star-shaped pentagons indicated in Figure 39.

3. Use the pentagons and star-shaped pentagons of Exercise 2 above to make a linear scale, taking AB as the unit, then AE as μAB, then AF as $\mu^2 AB$, AG as $\mu^3 AB$, and so on, verifying the relations $1+\mu = \mu^2$, $\mu+\mu^2 = \mu^3$, and so on, or using them to construct your scale. You can now test paintings or drawings to determine whether the proportions are in the golden ratio. (If you have to change your unit, which is likely, draw perpendiculars to the scale line at A, B, E, F, G, . . . and fit another line $A'B'E'F'G'$ . . . in, where $A'B'$ is the new unit, and A' lies on the perpendicular through A, B' lies on that through B, and so on.) You can only do this if $A'B' > AB$. Why? Compare your figure with Figure 40, which shows how Modulor scales are constructed.

4. Le Corbusier divided a flat rectangular surface up into tiles, rectangular and square, subdivided in the golden ratio. See whether you can make an interesting tile pattern this way.

5. Draw a regular hexagon (6-gon), and extend the sides so as to obtain a star-shaped hexagon. Use this design for the plan of a house of worship, using as many hexagons and star-shaped hexagons as you feel will make your design attractive.

·5· THE *OPTICS* OF EUCLID

In this chapter, as a prelude to a discussion of the *Elements* of Euclid, we shall discuss his *Optics*. Euclid wrote many works besides the great *Elements of Geometry*, which it is thought was composed in Alexandria about 300 B.C., but some of these have been lost for ever, some have been conserved in more or less fanciful Arabic versions, and others survived in Greek texts with many emendations and interpolations. Among the last category is the *Optics*. The one witness to the work being by Euclid is Proclus of Alexandria, a fifth-century A.D. philosopher who wrote a copious commentary on the first book of Euclid's *Elements*, and mentions the *Optics*. Proclus will appear in the next chapter.

As in the *Elements*, Euclid assembles what was known of optics, and this comprises the knowledge of the Pythagorean school, of Egypt, of Babylon, and of many other civilizations. When we discussed Vitruvius, it was mentioned that theatrical scenery must have produced the optical illusion of distance, and this fact occurs in the writings of the Greeks, Democritus and Anaxagoras.

But Euclid's *Optics* does not deal with perspective as we have discussed it in Chapter 2. Instead the *Optics* is subordinated to philosophical speculations and confused hypotheses on the nature of light, and is physiologically incorrect on the nature of sight. It was not until several generations after Euclid, with Archimedes, whose works on optics have vanished, then with Ptolemy, whose work was partly preserved in an Arabic translation, that the idea of refraction was properly understood. Finally Heron of Alexandria, who left a work on optics, and another on the measurement of angles by optical means, showed a proper understanding of the simple phenomena of optics.

Euclid wished to prove his propositions geometrically, and needed to relate the object to what is seen by the eye. The nature of vision is fundamental to such a discussion, and there were mixed ideas in his time, and for centuries later. A first conception dates back to Pythagoras, who lived between 600 and 500 B.C.

The eye is filled with luminous rays which strike objects, and produce the impression of sight by reflexion back to the eye. This view of sight was in accordance with the statement of Empedocles that the inside of the eye is formed of fire and water, and the outside of earth and air, so that the four fundamental elements of the universe were satisfactorily accounted for.

A second hypothesis on the nature of vision was that of the atomic school, founded around 500 B.C. by Democritus, amongst others, which thought of matter as empty space filled with atoms in perpetual movement, a view which tallies with our present one. It was believed that the image of an object was something substantial, composed of an infinity of corpuscles projected by the surface of the object, the shape being maintained until it reached the eye.

Vitruvius, in his *Ten Books on Architecture* (Chapter 1), is rather cautious when he discusses the nature of sight, remarking, as a practical architect, that in any case our vision often deceives us. Plato, in the *Timaeus*, has it all ways, and says:

When the light of day meets the visual current, then like meets like, joins with it, and in this union it forms a unique body following the direction of the eyes whenever the place where the light which arrives from the interior accords with that which comes from outside objects. This luminous body, which is affected in the same manner, because it is homogeneous, when it touches objects, or is touched, transmits their movements to the whole body, until the soul, and produces the sensation which we call sight.

Five centuries after Euclid, Theon of Alexandria, in a series of lectures on Euclid's *Optics*, still shares Euclid's views on rays emanating from the eye, and does not dispute Euclid's assertion that these rays have *gaps*, as it were, between them, so that the whole of an object is never seen at once. Theon cites the illustration of a needle lost on the ground, with many people searching for it, and never seeing it, and says that it seems to us that we see a complete object, because visual rays move very quickly, omitting nothing (that is *continuously*), without jumps.

Theon gives an interesting proof by analogy that light does not merely fall on the eye, but that something streams out to the object viewed. He remarks that the sense receptacles for hearing, smelling and tasting are all hollow, or concave, so that anything affecting them stays for some time, being trapped, more or less. But the eye, Theon says, is convex (he is referring to the pupil), and unless rays *emanated* from the eye, vision would not be possible. It was not known then, of course, that in fact the receptor for light is not the exterior membrane of the eye, but the interior, concave retina.

Euclid's *Optics* begins without a prologue, so that his views on the nature of vision have to be deduced from his various postulates, or assumptions. He assumes that the visual rays emanate from the eye, and move in straight lines, which diverge so as to include the object viewed. In his first Proposition, Euclid proves that no object can be viewed in its entirety at any given instant, and his proof assumes gaps between the visual rays emanating from the eye.

Let PQ be the magnitude looked at (Figure 45), B the eye from which the visual rays BP, BR, BS and BQ emanate. Then since the incident visual rays are propagated by divergence, they do not fall on the magnitude RS in a contiguous manner, and so, on the magnitude RS there are intervals where the visual rays do not fall. Consequently, one does not see the magnitude PQ simultaneously, but since visual rays move very quickly, it appears that one does.

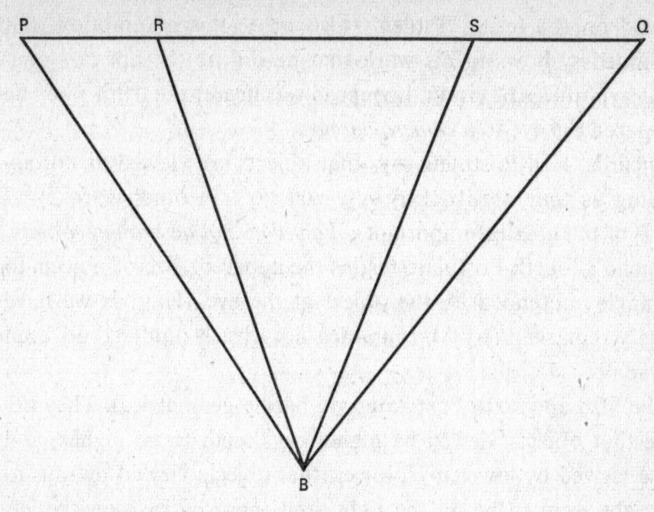

Figure 45

It is very odd indeed that Euclid should have postulated gaps in the visual rays which, he thought, emanate from the eye. We now know that light from a given object impinges on the retina, a thin nervous layer with a very complicated structure which covers the inside of the eyeball. The retina contains numerous receptor cells, the rods and cones, and these are stimulated by the light pattern which constitutes the retinal image. These cells, although very small and numerous, are finite in number, and there are minute gaps between them, so that one can say there are gaps in the light rays reaching the eye, as far as vision goes. So Euclid was right, but this may have been as lucky a guess as that of Democritus and his followers about the atomic nature of matter. In any case Ptolemy, in his second book on optics, adopts the postulate that visual rays are continuous, and not separated.

Euclid's second Postulate is that the visual rays form a cone around the contour of the object as base, and with the vertex of the cone in the interior of the eye. This hypothesis made it impossible to think of an image similar to the object viewed being formed on the retina. Ptolemy also made this assumption, but Archimedes shows in his work that he did not accept this one-point hypothesis of vision. Leonardo was nearer the truth when he compared the eye to a *camera obscura.*

Euclid's third Postulate says that objects are viewed or not according as they are reached by visual rays. In other words, and this is of the greatest importance, *light does not go around corners.*

Euclid's fourth Postulate relates the apparent size of objects to the angle subtended by the object at the eye. This, as we have already remarked (p. 88), is an idea not clearly understood even nowadays.

The fifth and sixth Postulates are hardly geometrical. They declare that objects viewed by higher rays seem to be higher, and those viewed by lower rays, lower, that objects viewed by rays to the right seem to be on the right, and those by rays on the left seem to be on the left! But the seventh Postulate is interesting. It says that *objects viewed by more numerous rays are more distinct.*

Let us see what this means. It is used in proving Proposition 2, which says: *among objects situated at a distance, those which are nearer are more distinct.*

Euclid takes two equal and parallel line segments PQ and RS (Figure 46) and uses the seventh Postulate. The angle RBS is larger than the angle PBQ, and therefore contains more diverging rays. By Postulate 7, it follows that RS looks more distinct than PQ.

This is possibly one of the simplest *proofs* that one could devise, in order to illustrate the idea of proof in mathematics.

Proposition 3 announces the physically true phenomenon that there is a distance beyond which a magnitude cannot be seen. The proof depends on the existence of a gap in the visual rays, and the reader is asked to prove this as an Exercise on p. 138.

Proposition 4 asserts that of two equal magnitudes, the one farther away looks smaller. The proof uses a fair amount of geometry, and is based on apparent size being related to the visual

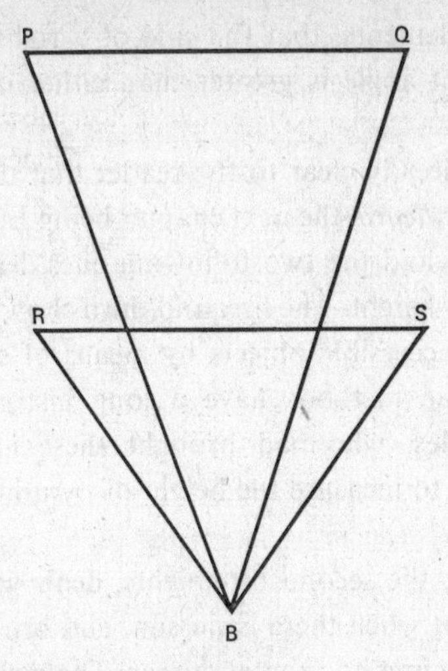

Figure 46

angle subtended at the eye by the object. We have already noted the apparent paradox mentioned by Leonardo da Vinci in his Notebooks, that columns farther away have to be painted larger on the canvas than nearer ones (p. 86). We saw that the visual angle subtended by a given circle at the eye *decreases* as the circle moves away from the eye, although the amount of circumference viewed *increases*, and that if we painted on a circular canvas, there would be no problem, since the depiction of the further circles would be smaller. If an object is depicted to be smaller than one expects it to be, it appears to be farther away, and this was well known to the ancient Greek and Roman architects.

Proposition 6 concerns the well-known phenomenon of the apparent convergence of parallel lines viewed from a point situated in their plane, or outside their plane.

Proposition 9 states that rectangular objects viewed from a distance look rounded, and again this proof is left to the reader. The proof is based on the Proposition that there is a distance beyond which a magnitude cannot be seen, and the Proposition, which appears in the Elements, that the side of a right-angled triangle opposite the right angle is greater than either of the other two sides.

Perhaps it is already clear to the reader that this chapter is an *Introduction and Allegro*, the next chapter being Euclid's Elements.

Proposition 18 and the two following ones deal correctly with three problems of height. The first and third show how to measure the heights of inaccessible objects by means of shadows cast on the ground. These methods have a long history. Even before Pythagoras, Thales, who had brought these ideas back from Egypt, knew how to measure the height of pyramids by the length of their shadows.

Proposition 19, the second on heights, deals with the problem of finding a height when there is no sun, and brings in the reflexion of a visual ray in a horizontal mirror. The method is shown in Figure 47, and depends on the angle of incidence, angle AHB,

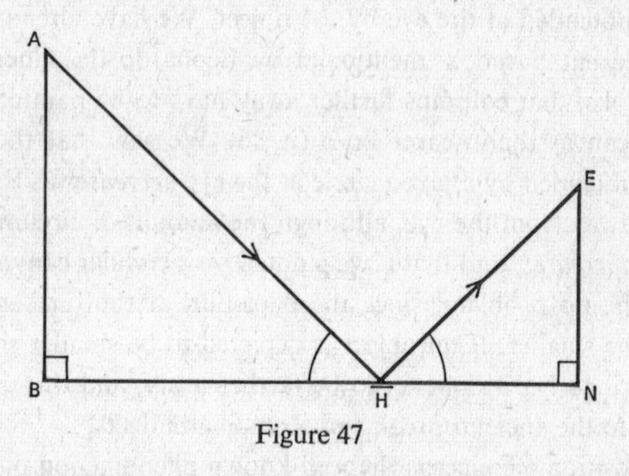

Figure 47

being equal to the angle of reflexion, angle EHN. The triangles ABH and ENH are similar, and so their corresponding sides are proportional. Hence

$$AB/BH = EN/NH,$$

so that if BH is known, and EN and HN are known, the height AB is known.

This sudden introduction of reflexions in a mirror puzzled the translators of the Renaissance, since reflexions and refractions are dealt with in another branch of Greek science, called *Catoptrics*, but they presumed that Euclid had written on this first, and so the first Latin versions of Euclid's *Catoptrics* and *Optics* put the *Catoptrics* first. But the *Catoptrics* does not deal with this kind of problem!

Propositions 34 and 35 are the most interesting geometrically, since they consider the aspect of the diameters of a circle when viewed from outside the plane of the circle. The question considered is: when do they all look equal, and when do they not? Pappus, at a much later date, considers the same Proposition, but gives a different proof, with no reference to Euclid, and concludes that in certain circumstances the circle looks like an ellipse. Euclid does show this, but expresses it differently, by saying in Proposition 36 that the wheels of a chariot sometimes look circular, and sometimes oblong!

The seven Propositions 50 to 56 have a kinematic character, since they deal with visual phenomena connected with magnitudes in movement. The first of these seven Propositions says that if the eye views a set of spaced-out magnitudes, moving parallel to each other, and with the same speed from left to right, the farthest one seems at first to move in front of the nearest one, but after passing in front of the observer, the farthest one seems to lag behind the nearest one. In Figure 48, BG seems to precede KA, but ST seems to precede NU. This is seen most easily by marking a number of points on the various lines in the diagram.

Proposition 51 is of historic interest, since it introduces for the first time the idea of *relative velocity*. If a magnitude moves at the same speed as the observer, it appears to be at rest. Proposition 54 deals with the idea of *motion parallax*. If magnitudes moving at the same speed are observed, those farther away seem to move more slowly than those which are nearer. This is of great practical importance, and used by pilots of aeroplanes when they are making a visual rather than an instrument landing.

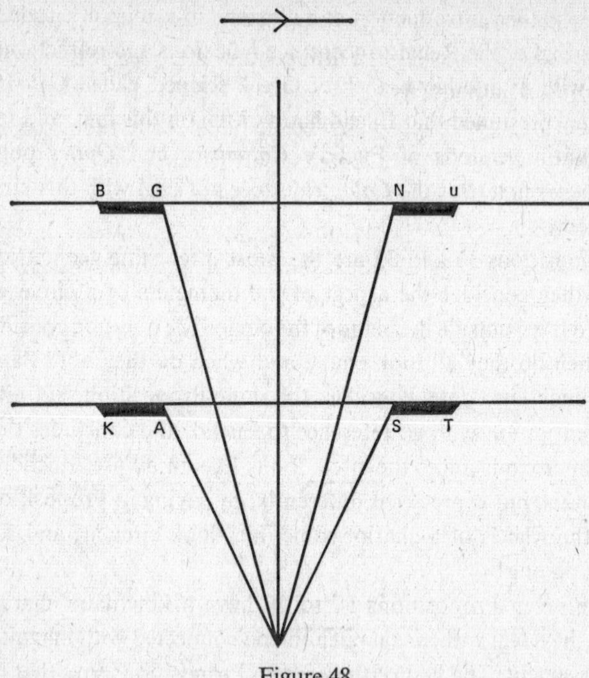

Figure 48

The further Propositions in the *Optics* are either of doubtful authenticity, or of no interest to us. Of the fifty-eight Propositions, twenty belong to optics as conceived nowadays, and the others to the theory of perspective, but as we remarked earlier, *nowhere does Euclid intersect his visual cone with a plane to see what happens.*

The *Optics* turned up very late in world history. It was not completely unknown during the Middle Ages, but it was overlooked by the zealous translators of the Renaissance. It had appeared in a Recension, constructed in the fourth century A.D. by Theon of Alexandria, but this differs considerably from the original.

The *Catoptrics* is geometrically inferior to the *Optics*, but it is the first work which deals with reflected and refracted light.

Lenses, however, are not mentioned. The work was first assigned to Euclid by Proclus, in the fifth century A.D., but it is considered of doubtful authenticity.

Some of the Propositions which deal with reflexions from mirrors confuse the *focal point* with the *centre of curvature*. We consider these two concepts.

The centre of curvature of a spherical mirror is simply the centre of the sphere of which the mirror is a part. Take a plane section through the mirror, and consider rays of light parallel to the axis of the mirror, and very near to the axis (Figure 49).

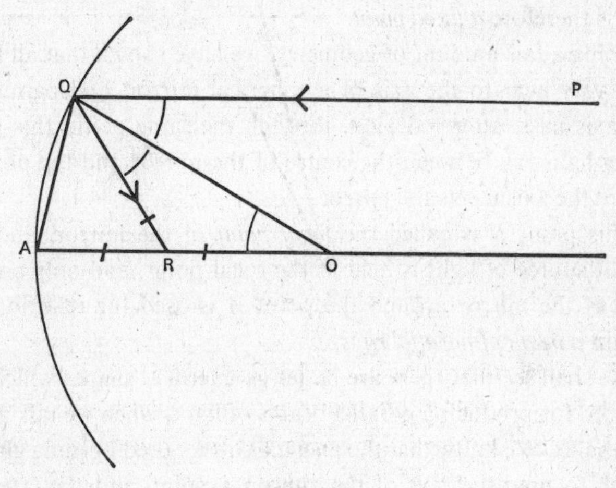

Figure 49

Let PQ be such a ray. Then the reflected ray QR cuts the axis at a point R where, by the laws of reflexion, known to Euclid, the angles PQO and OQR are equal, that is: *the angle of incidence is equal to the angle of reflexion*, OQ being the normal (perpendicular) to the surface of the mirror from which the angles are measured. Since PQ is parallel to OA, where OA is the axis of the mirror, the angles PQO and QOR (being *alternate* angles) are equal. We are using a Proposition of Euclid's *Elements* on parallel lines here, and this will appear again in the next chapter.

In the triangle OQR, the angles OQR and ROQ are equal, and hence the sides OR and RQ are equal (another Proposition of the *Elements*). If we join Q to A, the angle QAO is very nearly equal to a right angle, since Q is supposed to be very near to A, and since the angle QRA is equal to the sum of the angles RQO and QOR, and is therefore equal to twice the angle QOR, which is small when Q is near to A, the side RQ of the triangle QRA is nearly equal to the side RA. We therefore deduce that when Q is near to A, the point R is approximately *midway* between O and A, and is therefore *a fixed point.*

Using a fair amount of geometry, we have shown that all light-rays very near to the axis of a spherical mirror, and parallel to the axis pass, after reflexion, through the same point, this point being half-way between the centre of the mirror and the point A where the axis meets the mirror.

This point R is called *the focal point* of the mirror, and if a bright source of light is held at the focal point, and only a small part of the mirror around the point A is used for reflexion, we obtain *a beam of parallel rays.*

We shall see that there are better geometrical shapes which can be used for producing parallel beams of light, when we talk about parabolas. We know that the ancient Greeks used burning glasses, which focused the rays of the sun on a point, and the last Proposition of the *Catoptrics* does mention parallel rays being brought to a focus after reflexion in a concave mirror, but although the first part of the proof distinguishes between the focal point of the mirror and the centre, the second part confuses the two!

The *Catoptrics* deals only with the images of points, and does not contain any interesting Propositions about the images of objects. We conclude our discussion of the *Catoptrics* by looking at what is rather intriguingly called *Definition 6*:

If an object is placed inside a vessel at such a distance that it can no longer be seen, and if, this distance remaining the same, water is poured into the vessel, then the object will become visible.

In this context a Definition is a hypothesis. Figure 50 shows what happens, and this phenomenon is easily explained, using the

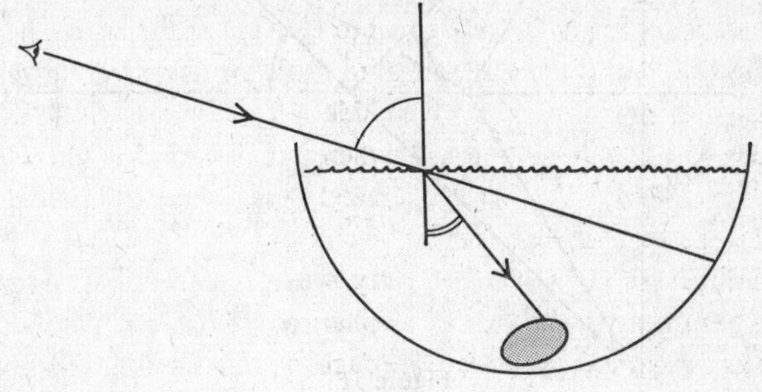

Figure 50

notion of *refraction*, the rays of light from the denser medium, water, being bent *away* from the normal to the water-surface at the points where they strike it, or, if we keep to the Greek view, the rays of light from the eye are bent *towards* the normal as they enter the water. No explanation of this kind appears in the *Catoptrics*.

It was Heron of Alexandria (*c.* A.D. 75) who explained the law of the reflexion of light by the *Principle of Least Time* (Figure 51). Let AB be a section of a plane mirror, and suppose that rays of light from P are being reflected by the mirror to pass through Q. Heron showed that if PMQ is the path of the reflected ray, then *the distance* PM+MQ *is less than the distance* PX+XQ, *where* X *is any point other than* M *of the mirror section* AB.

To prove this, we drop a perpendicular PN onto AB, and extend it to P', where $P'N = NP$. We join P' to Q, and suppose that this join cuts AB in the point M. Then PMQ is the path of the reflected ray, since the triangles PNM and $P'NM$ are con-

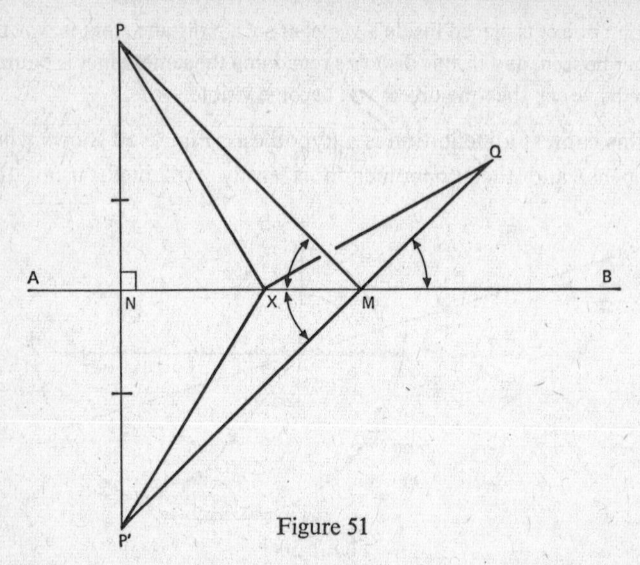

Figure 51

gruent, so that the angles PMN and $P'MN$ are equal, and the angle $P'MN$ is equal to the opposite angle QMB. Since the triangles PXN and $P'XN$ are also congruent, $PX = P'X$, and therefore $PX+XQ = P'X+XQ$. But we note that $P'X+XQ$ is the sum of two sides of the triangle $P'XQ$, and there is the Euclidean Proposition that *the sum of two sides of a triangle exceeds the third side*. Hence

$$P'X+XQ = PX+XQ > P'Q = P'M+MQ = PM+MQ,$$

and so we have proved that the path taken by the ray of light on reflexion is the shortest path, via the mirror, which is possible and therefore, assuming the speed of light to be uniform, the *time* taken to go from P to Q is the shortest possible. (This beautiful proof assumes a number of Propositions which appear in the next chapter.)

We conclude this chapter by describing the apparatus characteristically invented (or so we believe) by Leonardo da Vinci for solving a problem in optics of great antiquity. Here A and B are

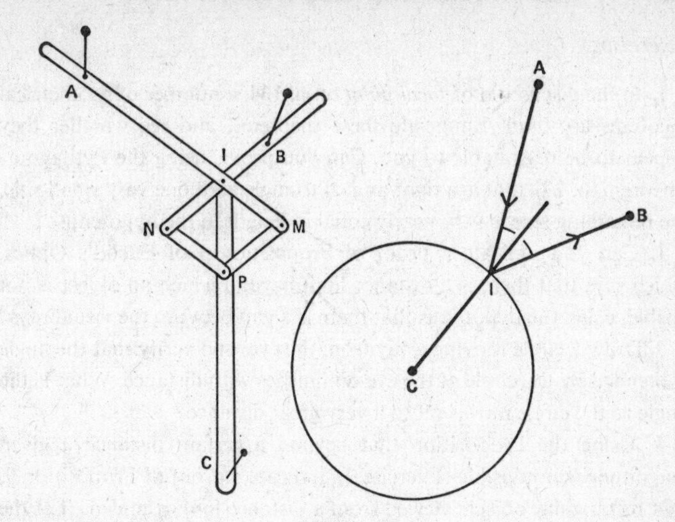

Figure 52

two given points in the plane of a circular mirror (Figure 52), and it is required to trace the path of a ray of light which starts at A, is reflected by the mirror, and passes through B. Let C be the centre of the given circular mirror, and let I be a point on the mirror where the ray AI is reflected to pass through B. Then since CI is the normal to the surface of the mirror, the laws of reflexion require that the angles CI produced makes with AI and with BI should be equal.

In the Leonardo apparatus there are joints at I, N, M and P, and $IM = MP = PN = NI$, so that the slot in which I and P move is always an axis of symmetry, and PI extended always bisects the angle AIB. The points A and B are fixed in the plane by pins, and we move the movable parts of the apparatus until the point I lies on the given circle, centre C. The path of the ray is then AIB. The theoretical solution to this problem involves the construction of an ellipse, with foci at A and B, touching the circle centre C. We shall refer to this again in Chapter 8, p. 208.

Exercises

1. In the discussion of *focal point* on p. 134, a number of geometrical theorems are used. Enunciate these theorems, and see whether they appear to be reasonable to you. Can you prove, using the Pythagoras Theorem (p. 27), that in a right-angled triangle with one very small side, the remaining side is very nearly equal in length to the hypotenuse?

2. Can you indicate a proof of Proposition 3 of Euclid's Optics, which says that there is a distance limit beyond which an object is not visible, using the hypothesis that there is a gap between the visual rays?

3. Draw a circle moving away from the eye, and verify that the angle subtended by the circle at the eye diminishes with distance. What is the angle as the circle moves off to a very great distance?

4. Using the Proposition that beyond a certain distance a given magnitude is not visible (Exercise 2), indicate a proof of Proposition 9, that rectangular objects viewed from a distance look rounded. (Let the rectangle be directly in front of you, and show that the corners are farther away than other points on the edges. Then use Proposition 3.)

5. In the problem of finding a height by means of a mirror (p. 130), if BH is 30 ft, HN is 5 ft, and NE is 2 ft, find the height AB.

6. Illustrate Proposition 50, that if the eye views a set of spaced-out magnitudes moving parallel to each other, and with the same speed from left to right, then the farther one seems at first to move in front of the nearer one, but this situation is reversed after the objects pass in front of the observer. (Give different positions of the moving objects.)

7. Discuss the notion of motion parallax, by plotting the positions of two moving objects, one nearer to an observer than the other.

8. Draw the path of several rays parallel to the axis and near to it for a spherical concave mirror, and verify the existence of a focal point on the axis through which the reflected rays pass approximately. This point should be half-way between the centre and the mirror.

9. If you have never witnessed the phenomenon of refraction described on p. 135, carry out the experiment, and draw your own diagram to show what happens as you pour water into the container.

10. Facing a slightly steamed-over mirror, trace the outline of your face in the mirror, keeping only one eye open. Verify that the outline is only one half the width and one half the length of your face.

Assuming that the image of your face is the same distance behind the mirror that your face is in front of the mirror, see if you can show, in a diagram, why the outline on the mirror surface as viewed by one eye is only half the size of your face.

·6·

EUCLID'S
ELEMENTS OF GEOMETRY

EUCLID'S *Elements* is, without any doubt, the most influential mathematics book ever written, and we shall try to explain why this is so. The first Latin author to mention Euclid is Cicero (106–43 B.C.), but it is not likely that in Cicero's time Euclid had been translated into Latin, or was studied to any considerable extent by the Romans; for, as Cicero also says, while Geometry was held in high honour among the Greeks, the Romans limited its scope by having regard only to its utility for measurements and calculations.

But by degrees the *Elements* did pass, even among the Romans, into the curriculum of a liberal education. During the Middle Ages the *Elements* survived, as did so much else, in Arabic translations, and it was not until 1482 that the first printed edition, in Latin, appeared in Venice.

This was the first printed mathematical book of any importance. It had a margin of two and a half inches, and the figures relating to the Propositions were placed in the margin. This method has recently been revived with certain lush calculus books.

The first, and most important, English translation, by Sir Henry Billingsley, appeared in London in 1570. It consists of 928 folio pages, not counting a long preface, and the figures for the Propositions in three-dimensional geometry are given twice, once in Euclid's version, and again with pieces of paper pasted at the edges so that the pieces can be turned up and made to show the real forms of the solid figures represented.

A bibliography, and no more, of the editions of Euclid would fill a large volume. Let us leave such considerations and, instead, turn to the *Elements*, and see how Euclid begins. He does so with *Definitions*, and we give some of them, numbered as in the *Elements*.

1. A *point* is that which has no part.
2. A *line* is breadthless length.
3. A *straight line* is a line which lies evenly with the points on itself.
7. A *plane surface* is a surface which lies evenly with the straight lines on itself.
8. A *plane angle* is the inclination to one another of two lines in a plane which meet one another and do not lie in a straight line.
9. And when the lines containing the angle are straight, the angle is called *rectilineal*.
10. When a straight line set up on a straight line makes the adjacent angles equal to one another, each of the equal angles is *right*, and the straight line standing on the other is called a *perpendicular* to that on which it stands.

We omit what follows until we arrive at

15. A *circle* is a plane figure contained by one line such that all the straight lines falling upon it from one point lying within the figure are equal to one another.
16. And the point is called the *centre* of the circle.
17. A *diameter* of the circle is any straight line drawn through the centre and terminated in both directions by the circumference of the circle, and such a straight line also bisects the circle.

The final Definition is:

23. *Parallel* straight lines are straight lines which, being in the same plane and being produced indefinitely in both directions, do not meet one another in either direction.

It is possible to discuss these Definitions at great length, and we can hardly leave them without any discussion, but we shall also bring in Proclus, one of the celebrated commentators on Euclid. But first let us give Euclid's *Postulates*. Euclid says:

Let the following be postulated:

1. To draw a straight line from any point to any point.
2. To produce a finite straight line continuously in a straight line.
3. To describe a circle with any centre and distance.
4. That all right angles are equal to one another.
5. That, if a straight line falling on two straight lines makes the interior angles on the same side less than two right angles, the two straight lines, if produced indefinitely, meet on that side on which are the angles less than two right angles.

There has been controversy over Euclid's Postulates, in particular the remarkable fifth Postulate, for centuries, and we illustrate in Figure 53 what Euclid intended to convey.

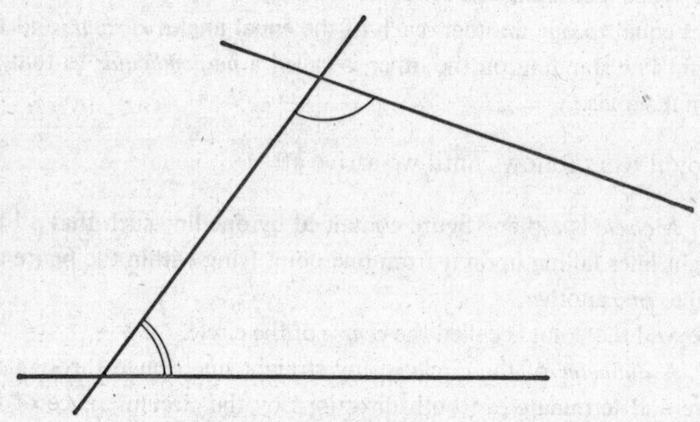

Figure 53

Euclid's introduction ends with some *Common Notions*:

1. Things which are equal to the same thing are also equal to one another.
2. If equals be added to equals, the wholes are equal.
3. If equals be substracted from equals, the remainders are equal.
4. Things which coincide with one another are equal to one another.
5. The whole is greater than the part.

There are many widely-used expressions in the English language which can be traced to the Bible and Shakespeare. Some also originate in Euclid's Common Notions. Once Euclid has prepared the ground, he goes ahead, and, in thirteen books, proves *Propositions* in geometry.

We have already indicated the great devotion to geometry manifested by Dürer, Leonardo and others. To Renaissance man the effect of reading Euclid was overwhelming. Beginning with a set of acceptable ideas, which we now call *axioms*, and a few constructions, Euclid was able to prove, using logic, each step being so evident that it was irrefutable, true relations between geometric objects, which he called Propositions, and we call Theorems, which, even if evident to a donkey in some cases (one of the criticisms), are almost beyond the power of belief in others.

Euclid proves that any two sides of a triangle added together exceed the third side. Put a donkey at one corner of a triangular fence, and some hay at another. The donkey will not go round two sides of the fence to reach the hay. On the other hand, the Theorem of Pythagoras, that the square on the hypotenuse of a right-angled triangle is equal to the sum of the squares on the other two sides, was not suspected by humans for thousands of years, and there are many theorems in Euclid which are as remarkable.

Many of Euclid's Definitions have been criticized. In order to avoid an infinite regress, definitions should not use ideas which, in their turn, also need definition. Dictionaries, which contain a finite number of words, avoid infinite regress by making their definitions *circular*. This can be verified by turning up the definitions of *point*, *line*, and so on.

For example, *point* is a place, or spot, a single *position*, and *position* is a place occupied by a *point*. If the reader wishes to have some fun, he can ask a friend what the term *point* means to him, and try to get a non-circular definition. He may find that frustration leads to abuse!

It is possible, as we shall see on p. 166, to overcome some of the difficulties inherent in a definition, but it may be thought that this is cheating, for we shall introduce points as *an undefined set of entities*, and lines as *certain subsets* of this set. However, let us return to Euclid.

It will be noticed that Definition 17 includes a theorem, that *the diameter of a circle bisects it*. Proclus, a fifth century A.D. commentator, ascribes this theorem to Thales, and remarks:*

> The cause of the bisection is the undeviating course of the straight line through the centre: for since it moves through the middle, and throughout all parts of its movement refrains from swerving to either side, it cuts off equal lengths of the circumference on both sides.

But Proclus adds:

> If you wish to demonstrate this mathematically, imagine the diameter drawn, and one part of the circle fitted upon the other. If it is not equal to the other, it will fall either inside or outside it, and in either case it will follow that a shorter line is equal to the longer. For all lines from the centre to the circle are equal.

The flavour of the first remark and that of the proof are entirely different. In the proof, Proclus indicates that a *mirror image* of one part of the circle be taken in the diameter, and that the image should be compared with the other part of the circle. The centre of the circle remains unchanged, and so any distinction between the two parts of the circle can only lead to the deduction that there are points on the circle nearer to, or farther from the centre than they can be. Hence the two parts must coincide, the part that is reflected and the other part. With some further indications as to what happens when a mirror image is formed, the proof would be complete.

The style is not pure Euclid, and we should consider the method of presentation which left its imprint on mathematics for cen-

*Morrow, *Proclus' Commentary on the First Book of Euclid's Elements*, p. 124.

turies. Proposition 1, *On a given finite straight line to construct an equilateral triangle*, will serve our purpose. The method of construction is one we have used already.

Let *AB* be the given finite straight line (Figure 54).

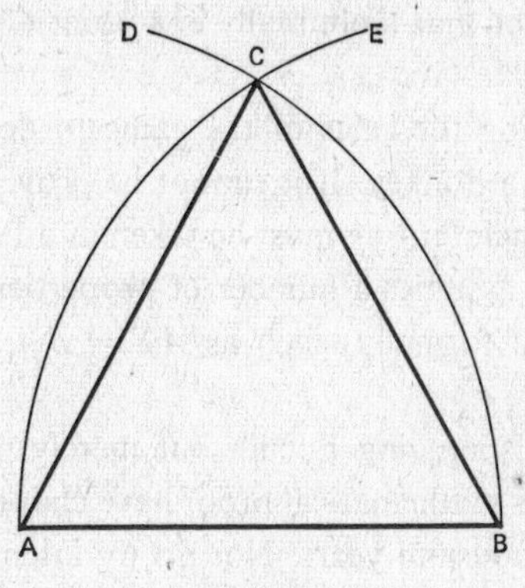

Figure 54

Thus it is required to construct an equilateral triangle on the straight line *AB*.
With centre *A* and distance *AB* let the circle *BCD* be described: [Post. 3],
again, with centre *B* and distance *BA* let the circle *ACE* be described:
and from the point *C* in which the circles cut one another, to the points *A*, *B* let the straight lines *CA*, *CB* be joined. [Post. 1].
Now, since the point *A* is the centre of the circle *CDB*,
AC is equal to *AB*. [Defn. 15].
Again, since the point *B* is the centre of the circle *CAE*,
BC is equal to *BA*. [Defn. 15].
But *CA* was also proved equal to *AB*; therefore each of the straight lines *CA*, *CB* is equal to *AB*.
And things which are equal to the same thing are also equal to one another: [C.N.1].
therefore *CA* is also equal to *CB*.
Therefore the three straight lines *CA*, *AB*, *BC* are equal to one another.

Therefore the triangle ABC is equilateral, and it has been constructed on the finite straight line AB.
[Being] what it was required to do.

Every step in the proof above appears to be justified by reference to Postulates, Definitions or Common Notions. But, . . . where is the proof that there really is a point C where the circles cut one another?

It is not playing the rules of the game to declare that this is *obvious*, even to a donkey. If it cannot be proved that the circles cut, the fact of their cutting must be taken as a Postulate.

Again, Euclid assumes a number of properties of length, without stating them formally, such as $AB = BA$. Is this also obvious?

We are not disparaging Euclid, but merely pointing out that attitudes towards mathematical proof have changed, as one might expect, in two thousand years. Nor do we intend to give a commentary on the thirteen books of the Elements. For this we must refer to the great *The Thirteen Books of Euclid's Elements*, by T. L. Heath.

Our aim is to give some indication of Euclid's style, and of the enormous effect that Euclid has had on people. Spinoza, the Dutch philosopher (1632–77) wrote his *Ethics* in the style of Euclid: Postulates, Common Notions and Propositions. Let us read what happened to Thomas Hobbes, the author of *Leviathan*, who was born in 1588 and died in 1679. The extract is from the delightful *Aubrey's Brief Lives*, edited by Oliver Lawson Dick.

He was 40 yeares old before he looked on Geometry; which happened accidentally. Being in a Gentleman's Library, Euclid's Elements lay open, and 'twas the 47 El. libri 1.* He read the Proposition. *By G* —, sayd he (he would now and then sweare an emphaticall Oath by way of emphasis) *this is impossible*! So he reads the Demonstration of it, which referred him back to such a Proposition; which proposition he

*The Theorem of Pythagoras (p. 27).

read. That referred him back to another, which he also read. *Et sic deinceps* [and so on] that at last he was demonstratively convinced of that trueth. This made him in love with Geometry.

Bertrand Russell, in his autobiography, describes the revelation that came to him when he first encountered Euclid at an early age. But there were others who were revolted by the assurance of the style. In a novel I read many years ago, *The Mighty Atom*, by a romantic writer, Marie Corelli, there was a little boy, brought up by agnostic (or worse) parents, who was taught Euclid instead of the Bible. He came to a bad end.

Euclid was so much a part of a liberal education that abuses were bound to creep in. Until modern times students had to memorize the Propositions, with their numbers, and the number of the Book in which they can be found, so that if one of the enlightened wrote: 'By Euclid 1, 17', his brethren knew he referred to the Proposition of Book 1: 'In any triangle two angles taken together in any manner are less than two right angles.'

This kind of teaching seemed to call for improvement, and the Mathematical Association of Great Britain was originally founded as an association for the improvement of geometrical teaching, and in particular for the specific purpose of replacing Euclid. Since then thousands of texts have appeared, all designed to replace Euclid, and the net result has not been at all satisfactory. In fact geometry at high school level in some countries is practically non-existent.

But curiously enough the inadequacies of Euclid's Postulates, from the modern point of view, led to one of the greatest revivals in mathematics. In the late nineteenth century David Hilbert of Goettingen applied his magic touch to the foundations of geometry, and developed, together with many other things, an adequate set of axioms for the development of Euclidean geometry. This adequate set of axioms is unfortunately too involved for satisfactory high school teaching, but Hilbert's work on the foundations of geometry established the fundamental importance of the axiomatic method in modern mathematics, and the enormous activity of twentieth-century mathematics is certainly grounded on the axiomatic method.

Since mathematics is a man-made activity, it is always wise to look back as well as forwards, in fact *se reculer pour mieux sauter*, and a close look at Euclid is rewarding.

We mentioned earlier that Euclid makes no provision for a compass-carried length. Proposition 2 is an interesting manifestation of this. It reads: *To place at a given point (as an extremity) a straight line equal to a given straight line.*

We have already given the construction, but, once again, it is of interest to read how it appears in Euclid.

Let A be the given point, and BC the given straight line [Figure 55]. Thus it is required to place at the point A (as an extremity) a straight line equal to the given straight line BC.

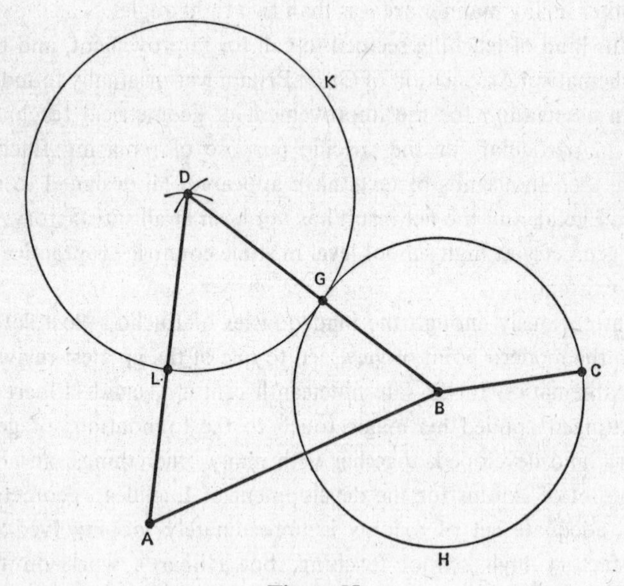

Figure 55

1. Caryatides in the room of the Swiss Guards in the Louvre

(Plates 1–9 are taken from the first French translation of Vitruvius's *Ten Books of Architecture*, published in 1684.)

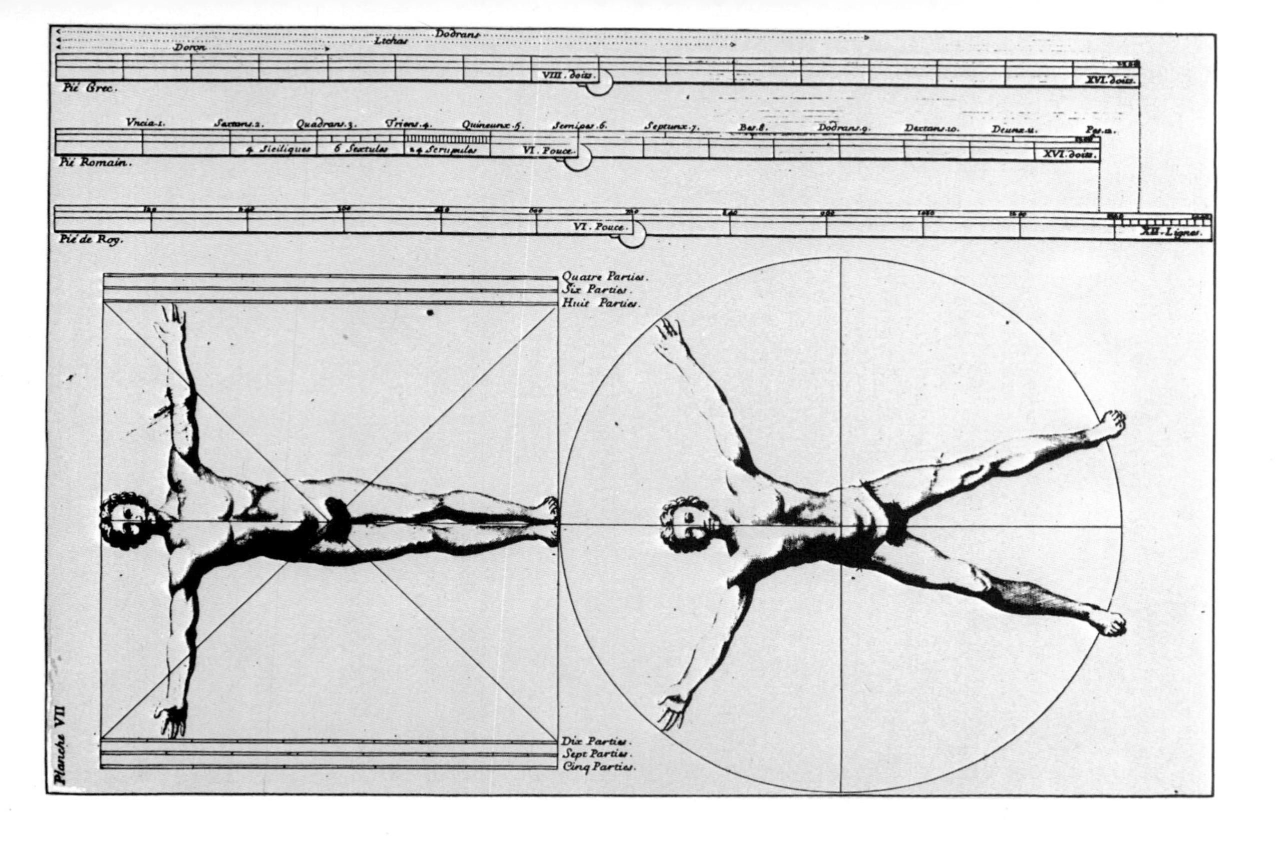
Dodrans
Lichas
Doron
VIII. doits.
XVI. doits.
Pié Grec.
Vncia.1.
Sextans.2.
Quadrans.3.
Triens.4.
Quincunx.5.
Semipes.6.
Septunx.7.
Bes.8.
Dodrans.9.
Dextans.10.
Deunx.11.
Pes.12.
4 Siciliques
6 Sextules
24 Scrupules
VI. Pouce.
XVI. doits.
Pié Romain.
VI. Pouce.
XII. Lignes.
Pié de Roy.
Quatre Parties.
Six Parties.
Huit Parties.
Dix Parties.
Sept Parties.
Cinq Parties.
Planche VII

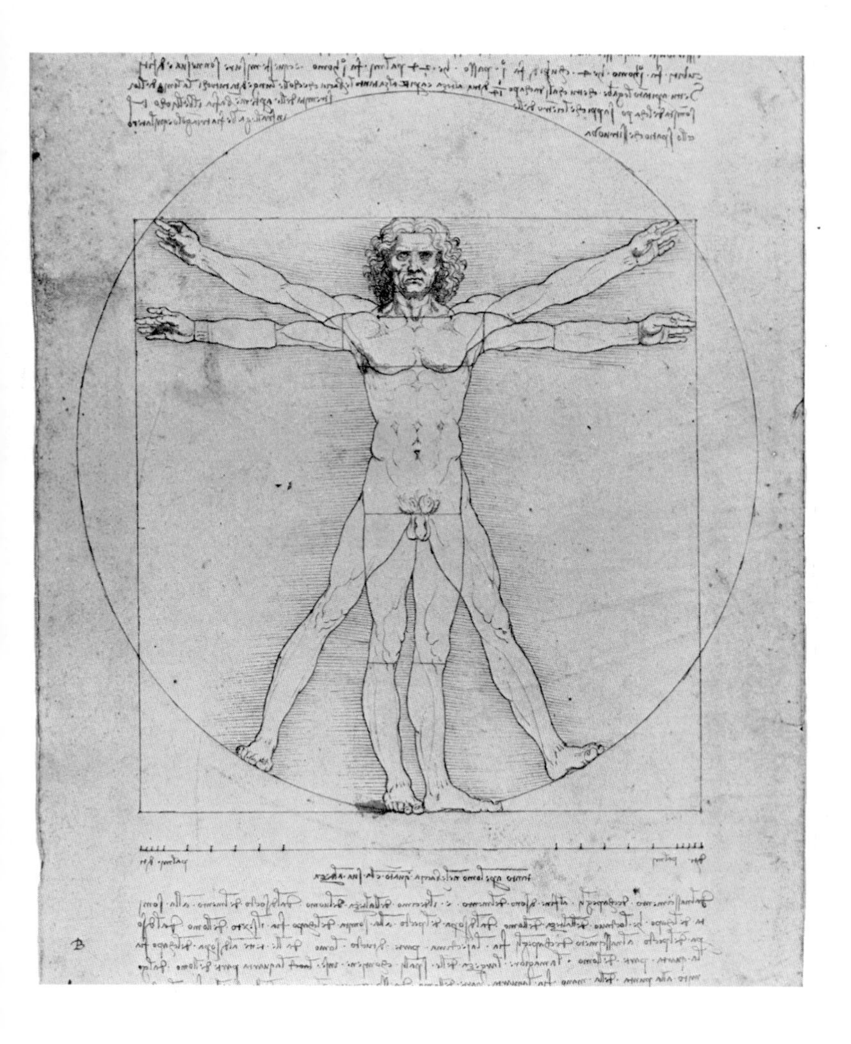

2, (opposite). A 1684 illustration of Vitruvius's concepts of human measurement. The scales at the side compare the Greek, Roman and Royal (French) units of measure.

3, (above). The influence of Vitruvius's ideas is clear in Leonardo da Vinci's famous drawing from his Notebooks.

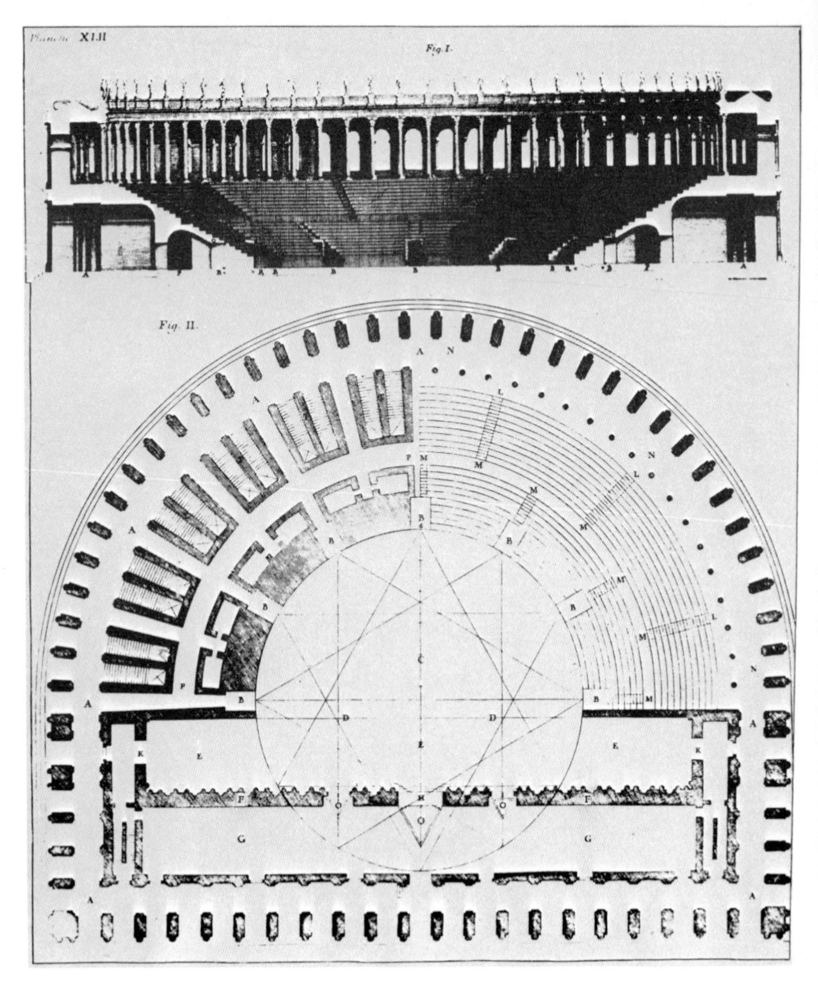

4 and 5. These designs of Roman (4, above) and Greek (5, opposite) theatres are based on different geometrical figures. Changes in plan followed developments in theatrical presentation.

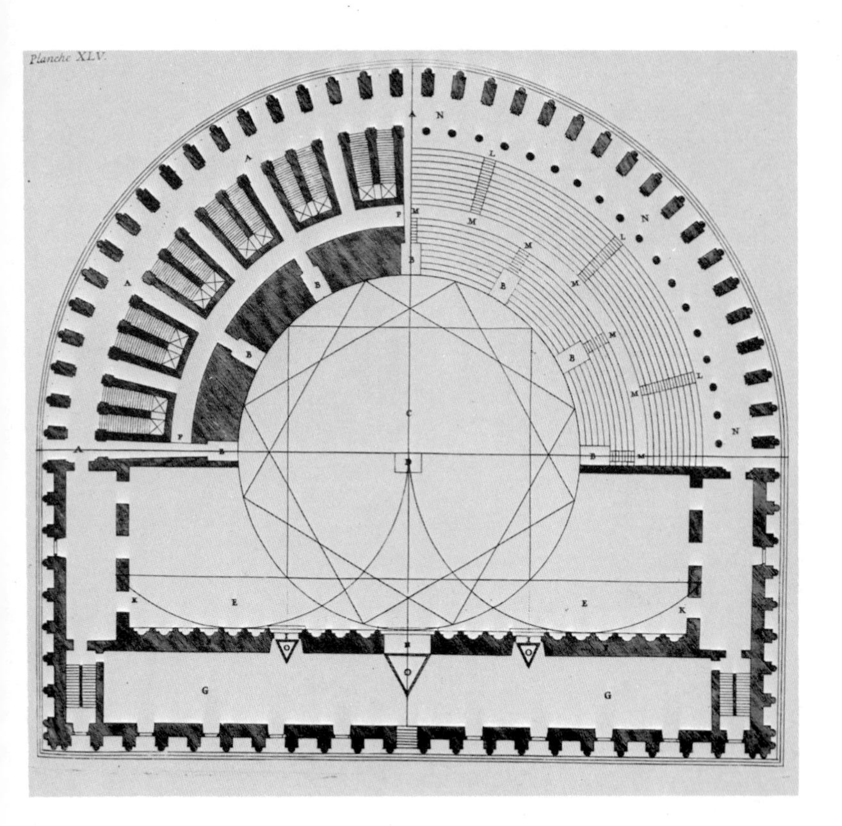
Planche XLV.
A
N
L
A
P
M
M
N
L
M
B
B
B
M
A
B
M
B
M
L
M
C
N
A
F
B
B
D
B
M
K
E
E
K
O
O
O
G
G

6. Vitruvius's version of the origin of the acanthus motif on the Corinthian column.

7. Designing war machines – catapults, *ballistae* and siege towers – was part of the Roman architect's duty.

Planche LVI.
Fig. II.
Fig. I.
A
B
C
D
E
F
K
L
M

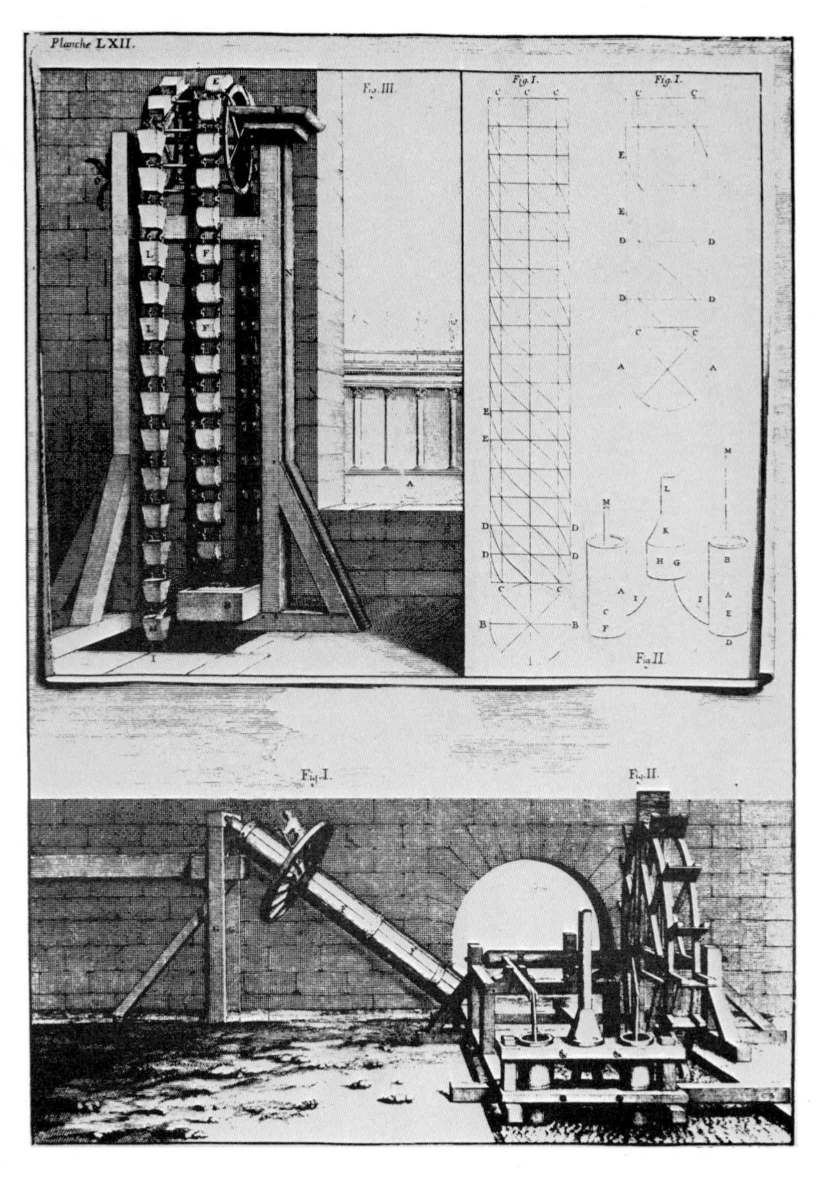

8 and 9. Water devices: a water clock (8) and an Archimedean screw (9, top right and bottom), both discussed by Vitruvius.

MELENCOLIA§I

10 and 11. *Melancholia I* and *St Jerome in his Cell* show Dürer's concern with perspective. Melancholy became associated during the Renaissance with the artistic temperament, or creativity, and Dürer's representation is surrounded by the attributes of geometry and the constructive arts.

12, 13 and 14 (overleaf). Some of Dürer's mechanized devices for representing perspective. In 12, above, and 13, below, the eye of the observer is fixed by a sight and, between it and the subject, a glass plate or a frame divided into squares by a net of black thread is interposed.

14. In this device the eye of the observer is replaced by one end of a string. The operator can then indicate various parts of the subject with the pointer at the other end, and mark them on the glass plate.

15. A pinhole camera photograph reproduces exactly what the eye sees. The correct perspective drawing of a sphere can vary from a circle to an ellipse, but artists have for centuries insisted on drawing circles, whatever the artist's viewpoint. (From M. H. Pirenne, *Optics, Painting and Photography*, C U P, 1970.)

16. Hogarth's comment on the importance of correct perspective, *False Perspectives*.

17. Where was the artist standing? In the foreground of Holbein's *The French Ambassadors* in the National Gallery, London, there is a curious pale object which turns out, viewed from the right position, to be a human skull.

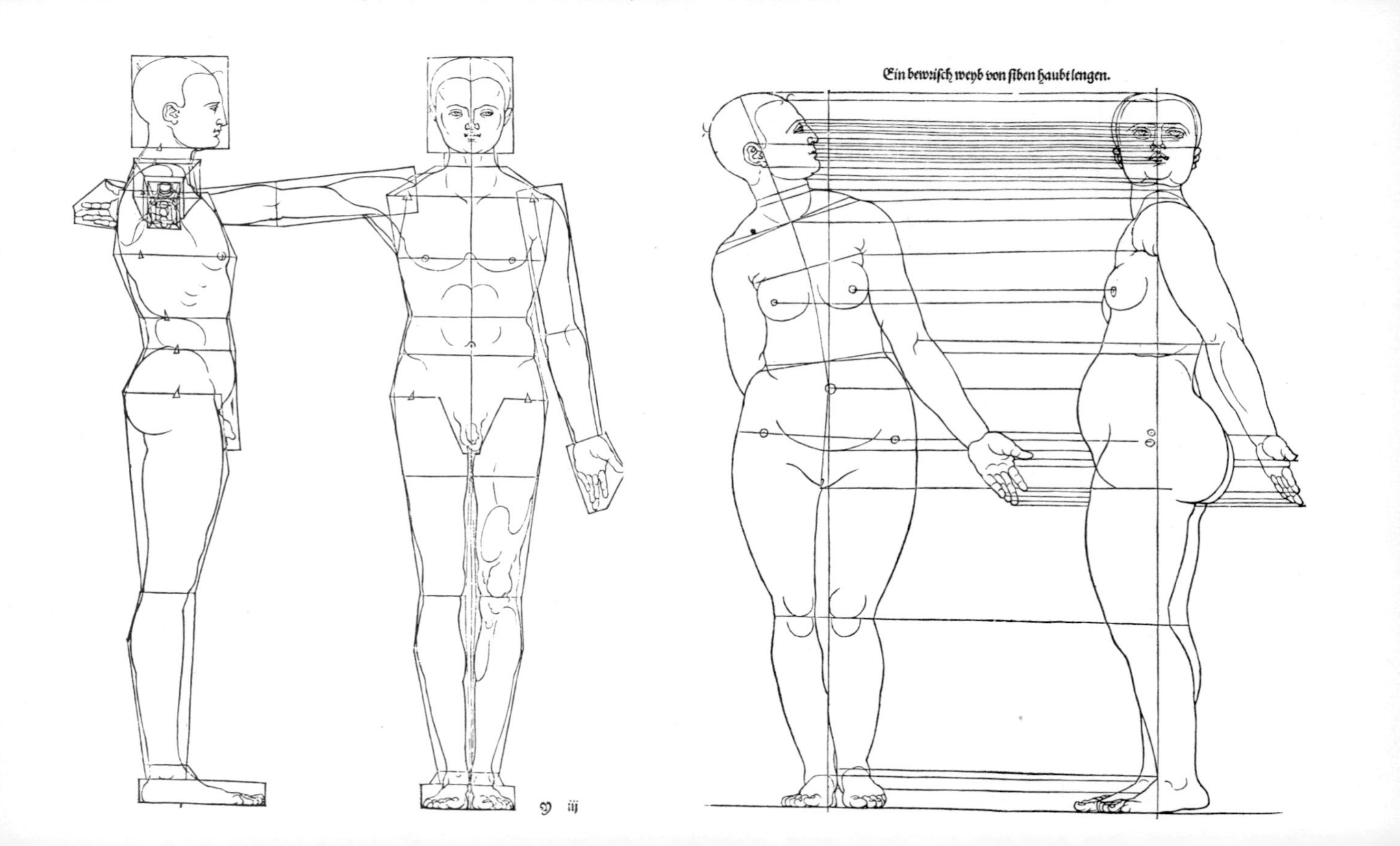
Ein bewrisch weyb von siben haubt lengen.

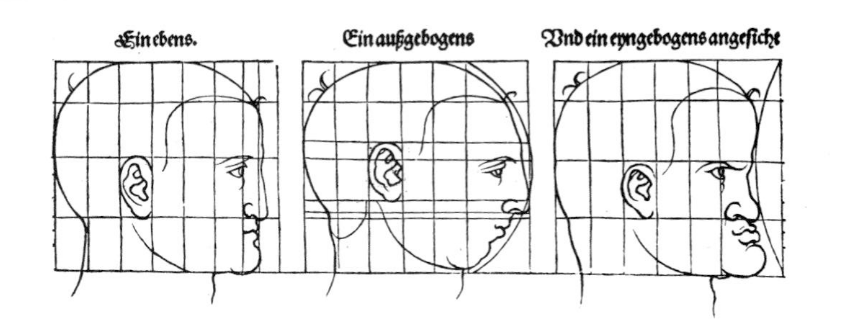

Es sind nach zweyerley fort der angesicht die da nach der seyten anzusehen inn irer vierung also zu rucken sind / Die zwo zwerch linien der vierung vnden vnd oben beleyben an beden vierungen recht zwerch linien die sichbarnn / doch müssen bede vierungen rautens weyß geruckt werdenn / das vernym also / Die erst vierung werdt mit irem fordern obern eck herfür / vnnd mit dem vndern hindern eck hinhinder gezogenn / Des gleychenn thut man dem widersins / Darnach teylt man durch die gestracktenn zwerch linien all vor beschrybne teyl wider ein / Aber so man die auffrechtenn wider eyn teylt / so werden mit sambt beden seyten eytel ort lini darauß / die da inn der ersten fürsich inn der andern hindersich hangen / Darnach zeucht man die gestalt linien des angesichts wider darein / dañ so finstu in beden vierungen was darauß wirdet / wie ich dañ dise angesicht zu negst nach den dreyen vorgemelten angesichten hab auffgerissen.

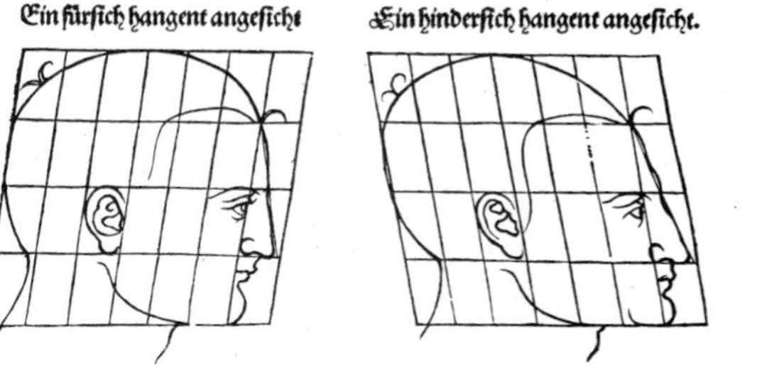

18, (above). Dürer's drawings in *Four Books of Human Proportions* make much use of geometry. Here, for example, the knee is found by bisecting the line joining the hip-point to the base of the figure. Although Dürer investigated the measurements of hundreds of people, he never pursued any studies of anatomy, unlike Leonardo.

19. Dürer described various methods, based on geometrical principles, of changing the proportions of human figures. This particular method had the effect of producing caricatures by geometry.

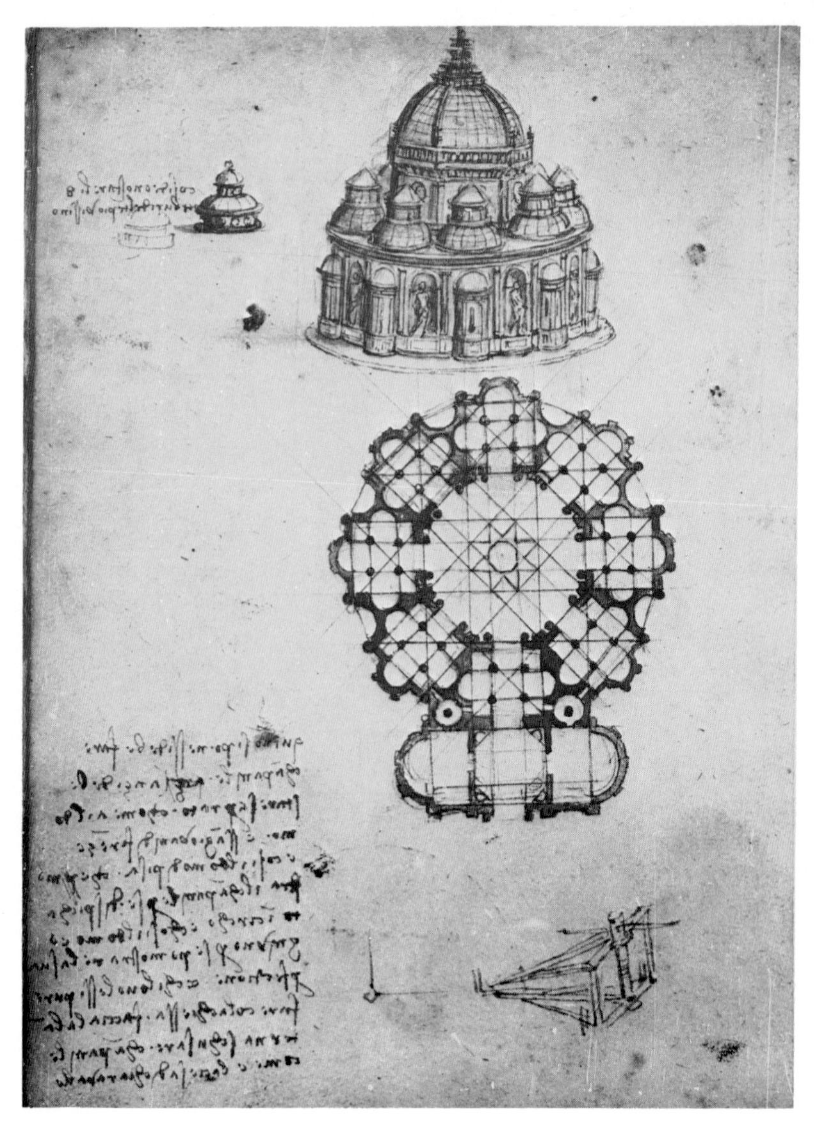

20, (above). Two of Da Vinci's plans for a church. Regular polygons are a fundamental part of the design. Da Vinci's interest in these geometrical figures is connected with his plans for adding chapels to churches without upsetting the symmetry of the main building.

21, (opposite). Like the rest of his Notebooks, this part of Leonardo's notes on *lunes* is written in mirror writing.

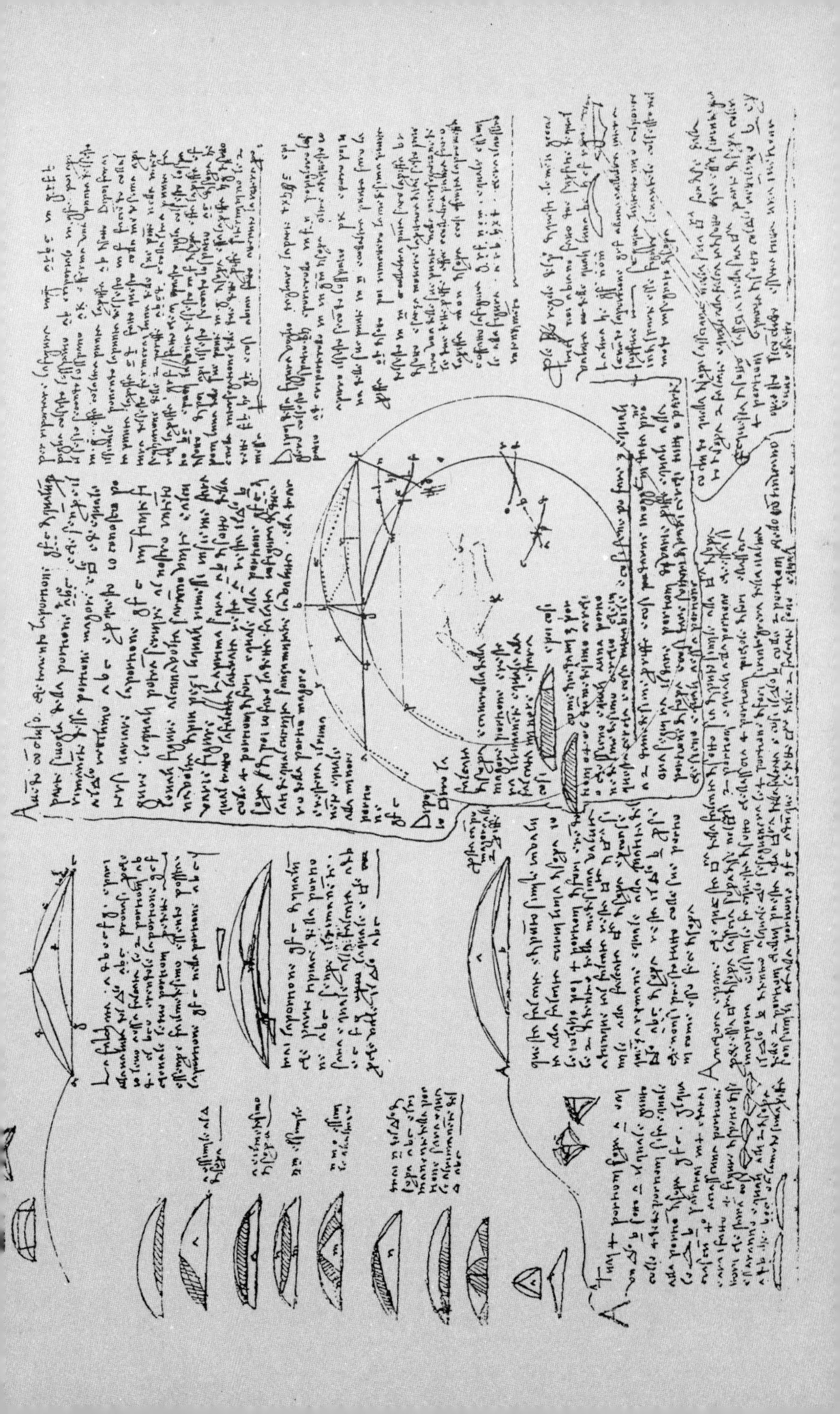

22. Repetition of similar geometrical figures and the use of the square superimposed on a double square in Renaissance architecture (Villa at Cesalto designed by Palladio).

23. The golden ratio used by Sir Theodore Cook in an analysis of a Botticelli Venus.

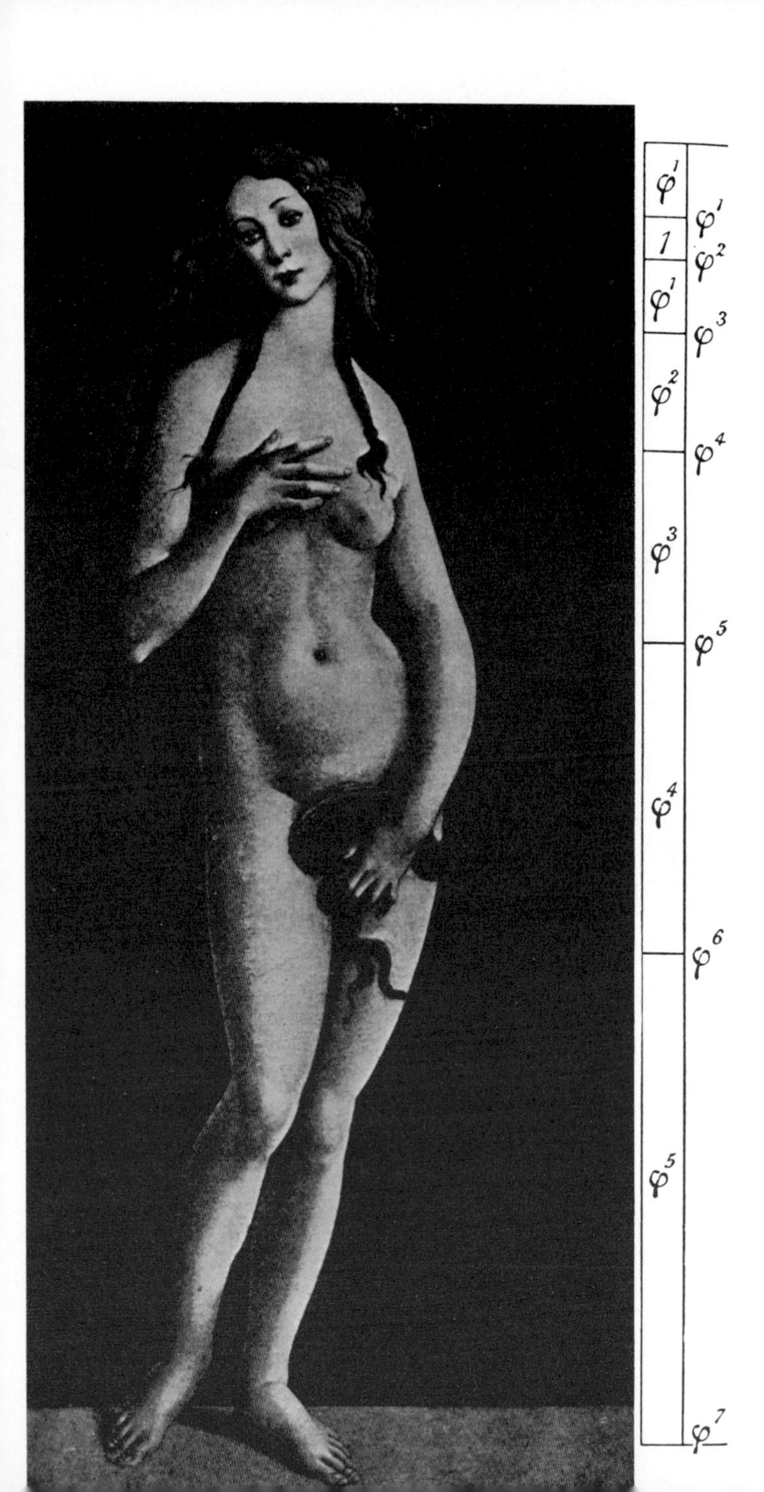
φ^1
1
φ^1
φ^2
φ^3
φ^4
φ^5
φ^1
φ^2
φ^3
φ^4
φ^5
φ^6
φ^7

24. Dodecahedron (on the table) and a semi-regular Archimedean solid in the portrait of Fra Luca Pacioli and his student Guidobaldo, Duke of Urbino, by Jacopo de Barbari (National Museum in Naples).

From the point A to the point B let the straight line AB be joined; [Post. 1]
and on it let the equilatéral triangle DAB be constructed [1, 1]
with centre B and distance BC let the circle CGH be described [Post. 3]
and again, with centre D and distance DG let the circle GKL be described [Post. 3]
Then, since the point B is the centre of the circle CGH,

BC is equal to BG.

Again, since the point D is the centre of the circle GKL,

DL is equal to DG.

And in these DA is equal to DB;
therefore the remainder AL is equal to the remainder BG [C.N. 3].
But BC was also proved equal to BG;
therefore each of the straight lines AL, BC is equal to BG.
And things which are equal to the same thing are also equal to one another [C.N. 1].
Therefore AL is also equal to BC.
Therefore at the given point A the straight line AL is placed equal to the straight line BC.
[Being] what it was required to do.

The fact that a previous proposition is used in the above proof makes it all the more exciting. The last sentence of a Proposition in Euclid is often left in the original Latin, *Quod erat demonstrandum*, which was to be proved, and the initials Q.E.D. have also taken their place in the English language as the clinching of an argument.

We shall examine just one more of Euclid's Propositions, to illustrate some very real difficulties in the axiomatic approach. It concerns what we call the Side-Angle-Side theorem, usually abbreviated to S-A-S theorem.

Proposition 4. *If two triangles have the two sides equal to two sides respectively, and have the angles contained by the equal straight lines equal, they will also have the base equal to the base, the triangle will be equal to the triangle, and the remaining angles will be equal to the remaining angles respectively, namely those which the equal sides subtend.*

Let ABC, DEF be the two triangles, having the two sides AB, AC equal to the two sides DE, DF respectively, namely AB to DE and AC to DF [Figure 56], and the angle BAC equal to the angle EDF.

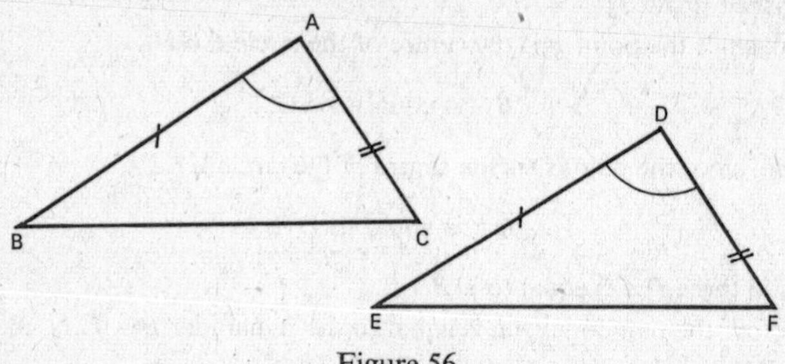

Figure 56

I say that the base BC is also equal to the base EF, the triangle ABC will be equal to the triangle DEF, and the remaining angles will be equal to the remaining angles respectively, namely those which the equal sides subtend, that is, the angle ABC to the angle DEF, and the angle ACB to the angle DFE.

For if the triangle ABC be applied to the triangle DEF, and if the point A be placed on the point D, and the straight line AB on DE,
then the point B will also coincide with E, because AB is equal to DE.
Again, AB coinciding with DE,
the straight line AC will also coincide with DF, because the angle BAC is equal to the angle EDF.
hence the point C will also coincide with the point F, because AC is again equal to DF.
But B also coincided with E;
hence the base BC will coincide with the base EF, and will be equal to it [C.N. 4].
Thus the whole triangle ABC will coincide with the whole triangle DEF, and will be equal to it,
and the angle ABC will equal the angle DEF,
and the angle ACB the angle DFE,

Q.E.D.

In this Proposition Euclid *applies* the triangle *ABC* to the triangle *DEF*. This is a matter which requires some explanation, but this is not given. A more serious objection is that the method works for the triangles *ABC*, *DEF* shown in the diagram, but fails for the triangles *ABC*, *DEF* shown in Figure 57. If two tri-

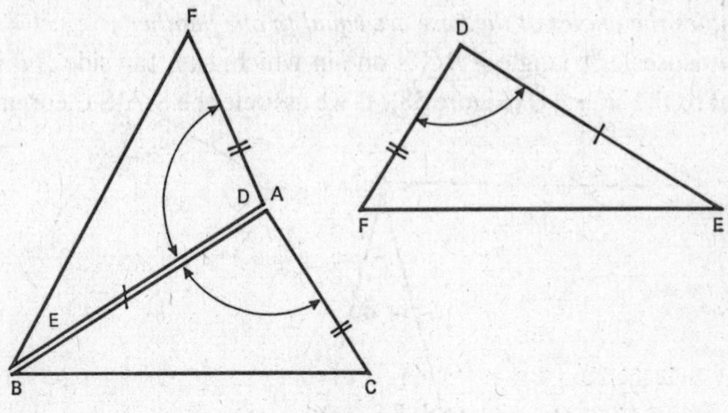

Figure 57

angles are mirror-images of each other, the method of superimposition does not work, and yet we feel that the triangles are identical. In fact, all that we have to do is to *lift triangle* DEF *out of the plane*, turn it over, and replace it, and we can then apply the method advocated by Euclid.

Again, although Euclid refers to the Common Notion 4: *Things which coincide with one another are equal to one another*, he is actually using a converse, and in fact, there are so many objections to Euclid's proof that many modern axiom systems, including Hilbert's, take the S-A-S theorem as an *axiom*.

The reader will feel uncomfortable if too many axioms are assumed in the course of a development of geometry. After all, why bother to prove anything? Why not take everything as an axiom? The whole point of mathematics, especially geometry, as we stressed earlier, is that many unsuspected results can be proved on the basis of a certain number of axioms being assumed, and it is this which makes mathematics exciting. Axioms, of course, must not contradict each other, and the fewer the better. We shall see later (p. 159) how great thinkers, for centuries, tried to eliminate *just one* of Euclid's Postulates, the fifth, and what the search led to.

We have already made use of Euclid's Proposition 5: *In isosceles triangles the angles at the base are equal to one another.*

An isosceles triangle ABC is one in which, say, the side AB is equal to the side AC (Figure 58). If we assume the S-A-S theorem,

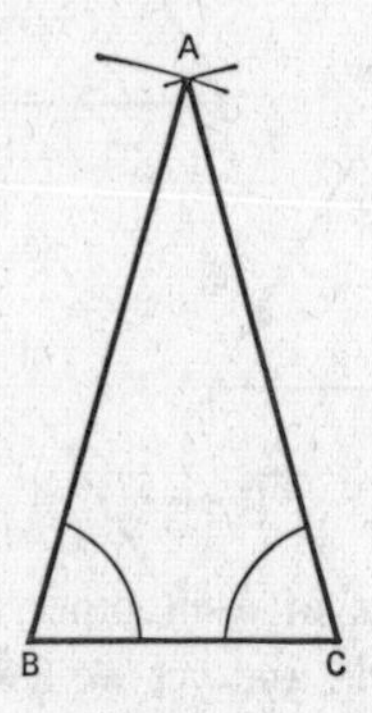

Figure 58

and compare the triangles ABC and ACB, we have $AB = AC$, $AC = AB$, and angle BAC = angle CAB, being the same angle, so that the triangles are *congruent* (this is the term we shall use, instead of *equal*), and therefore angle ACB = angle ABC, which is what we wished to prove.

Euclid's proof is longer, and the figure (Figure 59) has aroused interest for centuries. It has been called *pons asinorum, the bridge of asses*, and there is disagreement as to the reason! The figure, especially if drawn with the sides fairly flat, does suggest a trestle bridge, with a ramp at each end, and this could be traversed by a donkey, or a human being, but not by a horse. That is, a fool can get across easily, but the intelligent find it more difficult.

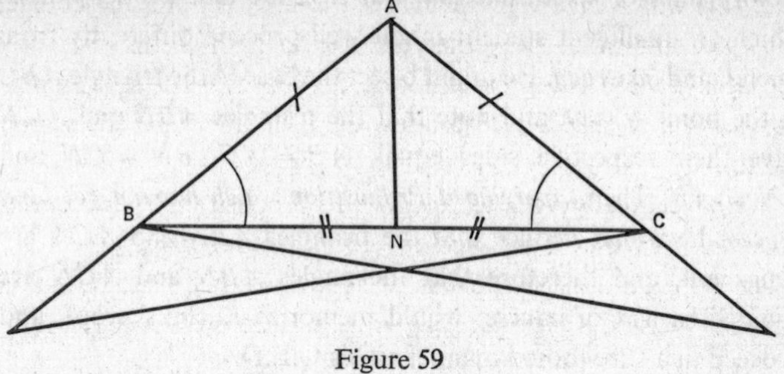

Figure 59

But the French reserve the term *pont aux ânes* for the figure of 1,47, Pythagoras' Theorem, the theorem which aroused Thomas Hobbes (Figure 60).

One hesitates to add another interpretation to the many contra-

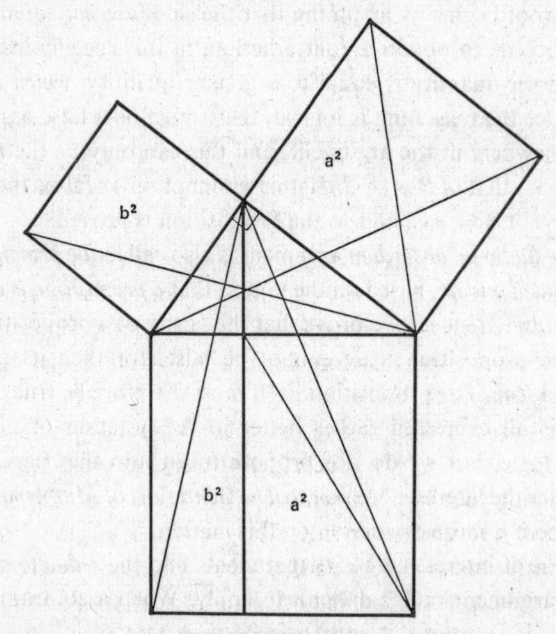

Figure 60

dictory ones of *pons asinorum*, but it is the first Proposition in which an intelligent student might well proceed differently from Euclid, and *go wrong*. He might bisect the base of the triangle ABC at the point N, say, and note that the triangles ABN and ACN have their respective sides equal, $AB = AC$, $BN = CN$ and $AN = AN$. Then, *applying a Proposition which has not yet been proved*, he would deduce that the triangles ABN and ACN are congruent, and therefore that the angles ABN and ACN are equal. The ass, of course, would memorize Euclid's proof, and proceed in a sure-footed manner to his Q.E.D.

It is in the converse to Proposition 1,5 that Euclid first introduces *reductio ad absurdum*, reduction to the absurd, and we must pause for a while to examine this notion. The converse to be proved states that if in a triangle ABC the angle ABC is equal to the angle ACB, then the sides AB and AC must be equal.

The proof begins by assuming that the sides are *not* equal, and then proceeds to obtain a contradiction, in this specific instance that a lesser quantity is equal to a greater quantity, *which is absurd.* Since the reasoning is logical, there must be a false assumption somewhere in the argument, and this can only be the *initial* hypothesis, that $AB \neq AC$. If this assumption is false, then we must have $AB = AC$, and so the Proposition is proved.

The *reductio ad absurdum* argument is also called *the principle of the excluded middle*, based on the notion that *a proposition is either true or false*. Hence if we prove that the falsity of a proposition is false, the proposition itself cannot be false, for then its falsity would be true. The proposition itself must therefore be true.

This is all expressed rather better in the notation of mathematical logic, but we do not propose to go into that here, and again refer the interested reader to *The Gentle Art of Mathematics*,* where there is more discussion on this matter.

What is of interest here is that students find the *reductio ad absurdum* argument rather difficult to apply. When a contradiction

*Pedoe, English University Press, 1965.

is reached, some students feel quite happy to stop there, not feeling that it is curious that there should be a contradiction. As Walt Whitman said:

> Do I contradict myself?
> Very well then I contradict myself,
> I am large, I contain multitudes.

But it must also be pointed out that some very eminent mathematicians regard the heavy reliance of mathematics on the *tertium non datur* as a pathological symptom, and refuse to accept any proof which uses it.

One of the most remarkable uses of the *reductio ad absurdum* is given by Euclid in proving that there is no rational number a/b, where a and b are integers, whose square is equal to 2. As we remarked, this discovery caused a crisis in Greek mathematics. Euclid proves this by *assuming* that *there is such a number*, and that we have cancelled all common factors of a and b. If $a^2/b^2 = 2$, then $2b^2 = a^2$, and so a^2 is *even*, divisible by 2. But odd integers, such as 1, 3, 5, . . . when squared are still odd, namely 1, 9, 25, . . . and so since a^2 is even, a must be even. Write $a = 2a'$, and square, then $a^2 = 4(a')^2$, and we have

$$2b^2 = 4(a')^2,$$

which leads to $b^2 = 2(a')^2$, and therefore b is even! But we cannot have both a and b even, since we cancelled all common factors. We have therefore reached a contradiction, and since our reasoning was immaculate, the original assumption must be wrong, and therefore $\sqrt{2}$ is *not* a rational number.

We now move back to the reconsideration of Postulate 5, which gives a sufficient condition for two lines *not* to be parallel. It is a remarkable fact that Euclid seems to shun this Postulate, and makes no use of it for the first twenty-eight Propositions. The geometry involved in these Propositions has been called *absolute geometry*, and there are geometries in which these Propositions continue to hold, *but Postulate 5 is denied.* Such geometries are called *non-Euclidean* geometries, and there are grounds for naming Euclid as the first non-Euclidean geometer!

Let us make a rapid survey of the theorems of absolute geometry. Some have been assumed in previous chapters. For instance, if one of the sides of a triangle is produced, the exterior angle is greater than either of the interior and opposite angles (Figure 61): angle ACD is greater than either angle CAB or angle ABC. (p. 109).

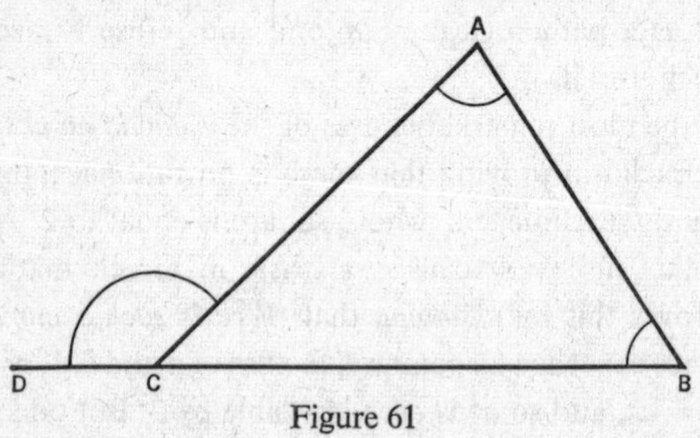

Figure 61

The theorem that the angles of a triangle add up to two right angles is a consequence of Postulate 5, and not a theorem of absolute geometry. We used this theorem on p. 71.

In any triangle the sum of any two sides exceeds the remaining side. This is a theorem of absolute geometry, so donkeys would continue to behave rationally in non-Euclidean geometry (p. 143).

Some of the theorems of absolute geometry give various criteria for two triangles to be congruent. They can be summarized as Side-Angle-Side, where the angle must be the included angle (already considered), Side-Angle-Angle and Side-Side-Side theorems. We must stress that although the equality of the respective sides of two triangles involves congruence, the mere equality of angles does not. Triangles with equal angles are called *similar* triangles, and their sides are proportional in length, which means that in Figure 62 the ratios $AB{:}GH$, $BC{:}HK$, $CA{:}KG$

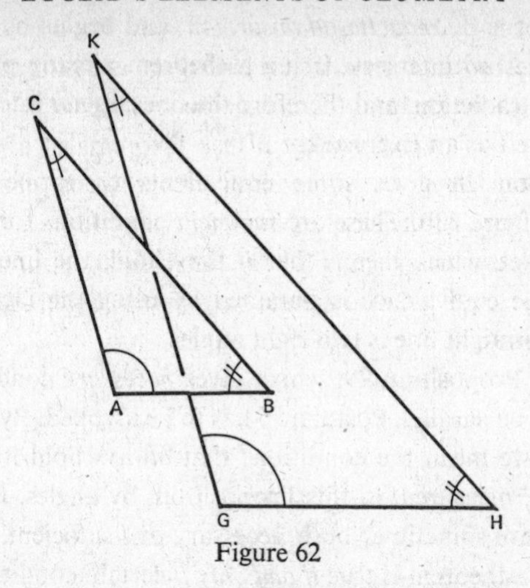

Figure 62

are all equal. On the other hand, there are non-Euclidean geometries in which all similar triangles are congruent.

Continuing with absolute geometry, in Proposition 27 Euclid proves that if a straight line falling on two other straight lines makes the alternate angles equal to one another (Figure 63a), then the straight lines will be parallel to one another.

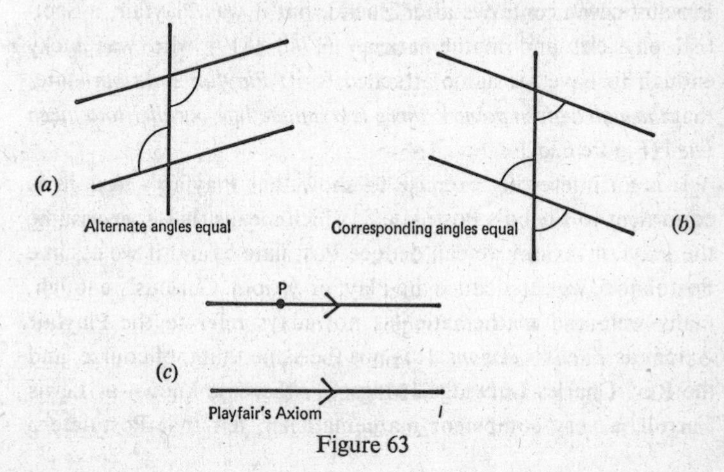

Figure 63

The proof is by *reductio ad absurdum*, and begins by assuming that the lines *do* intersect. Using a theorem already proved, we reach a contradiction, and therefore the lines do *not* intersect. The details are left as an Exercise (p. 171).

Proposition 28 gives some equivalents to Proposition 27, shown in Figure 63b. These are *sufficient* conditions for two lines to be parallel, which means that *if* they hold, the lines *must* be parallel. The equivalence is obtained by using the fact that the angle on a straight line is two right angles.

But with Proposition 29, which gives *necessary* conditions for two lines to be parallel, Postulate 5 has to be invoked. By *necessary* conditions we mean the conditions that *always* hold if two lines are parallel, measured, in this Proposition, by angles. Of course, conditions are sometimes both necessary *and* sufficient, and then we say that a theorem is true *if and only if* certain conditions hold. The conditions in Propositions 28 and 29 are necessary and sufficient conditions, but as we have pointed out, one proof is a proof in absolute geometry, and the other is not. Again, the reader is invited to explore a proof in an Exercise at the end of this chapter.

We now turn to Postulate 5. An equivalent form was known to Proclus, seven centuries after Euclid, but it was Playfair, a Scottish physicist and mathematician (1748–1819) who was lucky enough to have his name attached to it. *Playfair's Axiom* states that *through a given point* P *there is a unique line parallel to a given line* l (Figure 63c).

It is an interesting exercise to show that Playfair's Axiom is equivalent to Euclid's Postulate 5, which means that if we assume the Playfair Axiom we can deduce Postulate 5, and if we assume Postulate 5, we can deduce the Playfair Axiom. Curiously enough, many well-read mathematicians nowadays refer to the Playfair Axiom as *Euclid's Axiom*. It is not the same thing, of course, and the Rev. Charles Lutwidge Dodgson, otherwise known as Lewis Carroll, a very competent mathematician, felt that Postulate 5 was preferable, since you are given a very clear means of ascertaining whether two lines are parallel, whereas the Playfair Axiom is not helpful at all.

Many great minds have laboured to prove that Postulate 5 is *not independent* of the other Postulates, that *it can be deduced from them.* An indication of the work involved is given by the following partial list of *equivalents* to Postulate 5:

If a straight line intersect one of two parallels, it will intersect the other also (Proclus, fifth century A.D.);

There exists a triangle in which the sum of the three angles is equal to two right angles (Legendre, 1752–1833);

Given any three points not in a straight line, there exists a circle passing through them (Legendre, W. Bolyai, 1775–1856);

There exist two similar but non-congruent triangles (Saccheri, 1667–1733).

We have noted that Euclid had many critics and commentators, and the peculiar status of Postulate 5 attracted attention at a very early date. This interest led to the discovery of *non-Euclidean* geometry, one of the great leaps forward of the human mind. The apparent reluctance with which Postulate 5 is introduced by Euclid prompted geometers for over two thousand years to try to *deduce* Postulate 5 from the other Postulates. Proclus, to whom we shall return (p. 167), discusses an attempt by Ptolemy, which he shows is futile, and then gives a so-called proof of his own. The first remarkable published investigation, in 1733, was the work of an Italian Jesuit priest, Girolamo Saccheri. While he was Professor of Mathematics at the University of Pavia, he published a little book with the title: *Euclides ab omni naevo vindicatus* (Euclid Freed of Every Flaw). It appears that Saccheri had become very impressed with the power of *reductio ad absurdum*, and this method is applied to an investigation of the parallel postulate.

Suppose that $ABCD$ is a quadrilateral with right angles at A and B (Figure 64), and equal sides AD and BC. It is easy to show by absolute geometry that the angles D and C are equal.

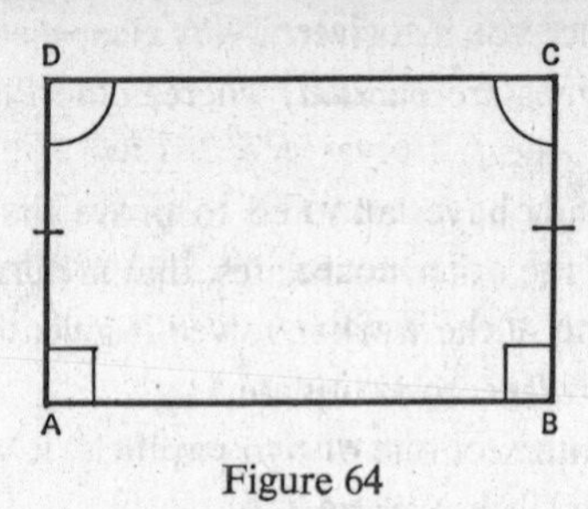

Figure 64

There are now three possibilities: the angles *D* and *C* are both *acute* angles (less than a right angle), the angles *D* and *C* are both *right* angles, or the angles *D* and *C* are both *obtuse* angles (greater than a right angle).

These three possibilities are referred to by Saccheri as: *the hypothesis of the acute angle*, *the hypothesis of the right angle*, and *the hypothesis of the obtuse angle*.

The plan of Saccheri's work is to show that both of the extreme hypotheses, of the acute angle and of the obtuse angle, lead to a contradiction. Then, by *reductio ad absurdum*, the hypothesis of the right angle must hold, and this hypothesis, as Saccheri shows, carries with it a proof of Postulate 5.

Tacitly assuming, as we must report Euclid also does, the infinitude of a straight line, Saccheri eliminates the case of the obtuse angle, but the hypothesis of the acute angle proves to be much more difficult. In his attempt, Saccheri obtains many of the theorems of what we now call non-Euclidean geometry, and if he had stopped there, he would have been credited as the first discoverer of non-Euclidean geometry. But he was so anxious to vindicate Euclid, that his reasoning ceases to be impeccable, and he finds an apparent contradiction.

It is now known that the geometry which can be developed under the hypothesis of the acute angle is as consistent as Euclidean geometry. In other words, the parallel postulate, Postulate 5, is *independent* of the other Postulates, and *cannot be deduced from them.* The three names associated with this great discovery are: Gauss (1777–1855) of Germany, often called *The Prince of Mathematicians,* Janos Bolyai (1802–1860) of Hungary, and Nicolai Ivanovich Lobachevsky (1793–1856) of Russia.

These three men approached the subject by considering a geometry in which the Playfair Axiom would be different, and considered the three possibilities: through a given point not on a given line *l* there can be drawn *more than one* line parallel to *l*, *just one* line parallel to *l*, and *no* lines parallel to *l*.

These three situations turn out to be equivalent, respectively, to the hypothesis of the acute angle, the hypothesis of the right angle, and the hypothesis of the obtuse angle. Again assuming the infinitude of a straight line, the third case was easily eliminated. Suspecting a consistent geometry under the first hypothesis, each of the three mathematicians independently carried out extensive geometric and trigonometric developments of the hypothesis of the acute angle.

Gauss did not publish his development of non-Euclidean geometry during his life-time, fearing the criticism of the unlearned and half-learned. Unfortunately Gauss withheld some much-needed praise when Bolyai's father, who had been a student, with Gauss, at Goettingen, acquainted Gauss with his son's work. Gauss's remark was that if he praised the young Bolyai, he would be praising himself, since he, Gauss, had already discovered the independence of Postulate 5!

It is now known that Gauss certainly discovered the possibilities of non-Euclidean geometry first, but Bolyai and Lobachevsky published their work at about the same time, in the 1830s, and share the credit for destroying a prejudice which had lasted for over two thousand years.

Why was the great Gauss reluctant to publish his discoveries? Immanuel Kant (1724–1804) had declared that there could be no other geometry than Euclidean geometry, that although its content comes *a posteriori* from sense perception, its form is determined by *a priori* categories of the mind. The influence of Kant was enormous, and the history of science shows that fixed ideas become very fixed indeed. Until recently there was a philosopher at the University of Oxford, always known as the home of lost causes, who became incensed whenever he met a young man rash enough to mention the possibility of non-Euclidean geometries.*

Let us see how we can persuade most people that non-Euclidean geometries do exist. We set up *a model* for a non-Euclidean geometry.

This is a miniature universe in which, with a suitable definition of points, lines, angles, and so on, all the first twenty-eight Propositions of Euclid are seen to be valid, but *the parallel axiom is denied.*

We shall not be able to give all the details, but it will be seen that this model is consistent, that is, free of internal contradictions, *if ordinary Euclidean geometry is consistent*, and we do not ask for more.

Our universe is the inside of a circle (Figure 65), our points are ordinary points inside the circle, but our *lines* are *the arcs of circles inside the given circle which cut the given circle at right angles.*

Let us call the inside Ω, and the circumference of our little universe ω. The first objection you may raise is that our lines are not *straight* lines. The answer to this is that Euclid only uses one property of straightness, which is that through any two distinct points P and Q a *unique* line can be drawn. Now it is true that through any two given distinct points P and Q inside ω a unique circle orthogonal (at right angles) to ω can be drawn.

(The proof of this theorem involves some knowledge of circle geometry, and we merely remark here that if O is the centre of ω, and we find a point P' on OP such that the product of lengths

*The elder Bolyai, in a letter written in 1820, warned his son against any obsessive interest in Postulate 5. 'You should detest it just as much as lewd intercourse, it can deprive you of all your leisure, your health, your rest, and the whole happiness of your life.'

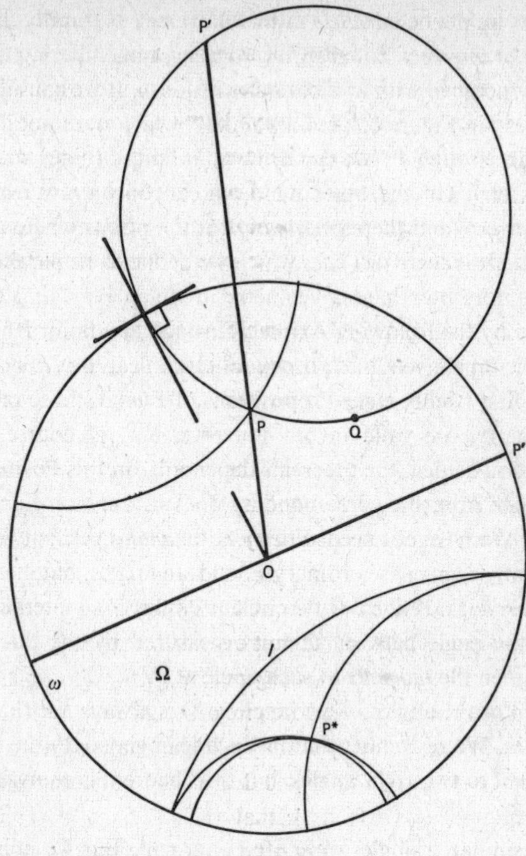

Figure 65

$OP.OP'$ is equal to the square of the radius of ω, then the circle which can be drawn through P, P' and Q is the unique circle through P and Q which cuts ω orthogonally. It will be noted that, because of this theorem, if two of our orthogonal circles intersect inside ω at a point P, they will also intersect *outside* ω at the point P'. Two of our new lines therefore intersect in *at most* one point, since points outside ω do not exist, for us.)

How do we define *parallels* in this universe? Naturally, as *lines which do not intersect, however far produced*, remembering that we are only concerned with arcs of circles inside ω. If we consider one of our lines l in Figure 65, and a point P^* which does not lie on l, we see that through P^* we can draw an infinity of lines which do not intersect l. On the other hand we can only draw *two* lines which intersect (and therefore *touch*) l at the points where l intersects ω, and it is *these two lines* which we define to be parallel to l.

We therefore now have a geometry in which Playfair's Axiom is replaced by the following Axiom: Through any point P^* which does not lie on a given line l, *two* parallels to l can be drawn.

All the first twenty-eight Propositions of Euclid, those of absolute geometry, are valid in our universe, but, of course, since Postulate 5 is denied, the theorems depending on this Postulate in Euclid differ from the corresponding ones in our non-Euclidean geometry. We have not needed to mention *angles* yet, but we now state that angles in ω are ordinary Euclidean angles, but measured between *circles*, not lines. If two of our circles in ω intersect at a point P, the angle between them is measured by the Euclidean angle between the *tangents* to each circle at P.

The sum of the angles of a triangle in ω is always *less* than two right angles. We remember that in Euclidean geometry the sum is always *equal* to two right angles, but this theorem is equivalent to Postulate 5.

Again, similar triangles in ω are congruent. But we must add immediately that this does *not* mean that one triangle can be superimposed on the other. Rigid motions (those which *preserve distance*) do have a counterpart in ω, but the explanation would take us too far afield. It is perhaps sufficient to state that *distance* can be introduced into ω, and when this is done the boundary ω of the universe is just as unattainable to the inhabitants of Ω as infinity is to us in our universe.

It was the great French mathematician Poincaré who invented the model we have just described, for a non-Euclidean geometry which fits in with the hypothesis of the acute angle. Poincaré also postulated physical conditions in this miniature universe which ensured that the steps taken by the inhabitants vary as they approach ω, so that the line joining P to Q is the shortest distance between P and Q, and as Q moves towards ω, the distance PQ tends towards infinity. To the inhabitants of Ω the geometry of their universe appears as necessary as Euclidean geometry appeared to Kant. It is fairly certain that Poincaré knew the quotation from *Hamlet*:

> I could be bounded in a nutshell
> And count myself a king of infinite space
> Were it not that I have bad dreams.

There are, of course, other models of non-Euclidean geometries, and the reader may have already thought of *the surface of a sphere*. On the surface of a sphere we are going to *identify* all pairs of points (P, Q) whose join passes through the centre O of the sphere (Figure 66). That is, a *point* in our new geometry is *a pair* of diametrically opposed Euclidean points on the surface of the sphere. A *line* is defined to be *the great circle* cut out on the surface by a plane through the centre O.

Then, through any two distinct points on the surface of the sphere there passes a unique line (based on the theorem that a unique plane can be drawn to pass through three given distinct non-collinear Euclidean points). Two distinct lines *always* intersect in this new geometry, and the intersection can only be one point. This is because two planes through O always intersect in a line through O, and this line cuts the surface of the sphere in a pair of diametrically opposed Euclidean points, that is in *one point* of our new geometry.

Hence in this geometry there are *no* parallels through a given point to a given line, and so this geometry corresponds to the hypothesis of the obtuse angle.

There is nothing in our discussion of the above non-Euclidean geometries which could not have been understood by Dürer, Leonardo or Saccheri, technically speaking. The enlarged concepts of point and line would probably have given them a lot of trouble.

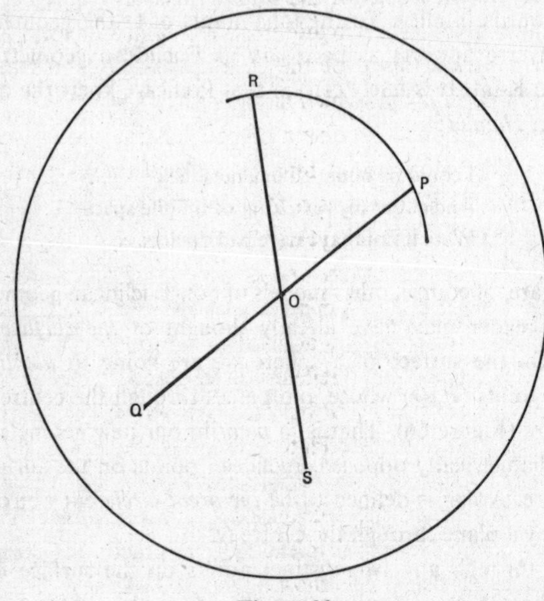

Figure 66

We mentioned Euclid's curious definition of a point, as that which has no part. The Pythagoreans, according to Proclus, defined a point as *a monad having position*. When we come to analytical geometry (p. 173), we may define a point in a plane as *an ordered pair of numbers* (p, q), but, of course, we shall have to know what we mean by numbers.

In modern geometry we begin with an undefined set of objects, which we call *points*, and a number of subsets of these points which we call *lines*. We say that a point is *on* a line, or *incident* with it, if the point is a member of a particular subset which denotes a line.

We then proceed from there, adding axioms appropriate to the kind of geometry we are investigating. But there are difficulties in the notion of a set, for which we refer again to *The Gentle Art of Mathematics* (p. 130).

In a very recent book on the foundations of geometry, Prof. Karl Menger gives an axiomatic treatment which enables him to revive, at a certain stage of the development, Euclid's famous definition: a point is that which has no part!

We have mentioned Proclus a number of times, and we shall now talk rather more about him. His commentary on the First Book of Euclid's Elements is a fifth century A.D. textbook on a work already more than 700 years old. Proclus, although not himself a creative mathematician, was an acute and competent critic, and immensely learned, his mind stored with the achievements of the preceding thousand years of mathematical inquiry.

Living at the end of the Hellenistic age, he could survey the accomplishments and errors of his predecessors with an Olympian eye. He was also a devout Platonist, and we shall see that this gives a very definite flavour to his remarks. He wrote at a time when Christianity was the official religion of the Roman Empire, but pagan beliefs were still tolerated. The practise of cults was forbidden, but Proclus, like most of the philosophers of Athens, was still attached to the ancient faiths, kept the sacred days of the Egyptians as well as those of the Greeks, and devoted himself assiduously to the study of mythological and religious observances.*

He taught at the famous Academy, still supported by endowments eight hundred years after Plato's death. We should give a lot to know whether the command: 'Let no one without geometry enter here!' was still carved on the portals. Proclus certainly knew a lot of geometry. He had enormous energy. Besides conducting five or six classes, holding other conferences and discharging his religious observances, he customarily wrote seven hundred or more lines a day!

*Morrow, Preface, p. vii.

The Commentary on Euclid is obviously a teaching text. It contains exhortations to students to work out various cases of the theorems discussed, and the mathematical reasoning and exposition is often interrupted by digressions pointing out the moral and metaphysical significance of a theorem, and its value in directing the mind upwards to the region beyond mathematical existence. This deep religious concern is an inseparable part of his, as of all Neoplatonic metaphysics.

Proclus regards Euclid as having been a member of the ancient Platonic school, and considers the ultimate purpose of the Elements to be the construction of the five *cosmic* elements, the five regular solids which figure so prominently in Plato's *Timaeus* (p. 261), and with which, in fact, the Elements conclude. But large parts of Euclid have no possible connection with this aim.

Proclus, as a good commentator, explains the meaning of the term *elements*, used as a title by Euclid. This means, of course, *beginning at the beginning*, but this is not understood by everyone nowadays. Some years ago I was informed by a professor of Latin, no less, that a book I had written could not be called *research*, and therefore included in a list of research publications, because I had said in the Preface: 'the treatment in this book is an elementary one'! As everyone who has opened a book on mathematics knows, the first page can be comprehensible, the treatment being elementary, but the development can contain a lot of research!

Sherlock Holmes's famous phrase: 'Elementary, my dear Watson!' is very apt in this context. Holmes would begin at the beginning, with the observation of a fact which had escaped the good Dr Watson, and then, by a series of brilliant logical deductions, lead up to the proposition which had astonished the doctor.

To return to Proclus – he is critical of many of the aspects of Euclid which we criticize nowadays. Euclid's definition of a plane angle as *the inclination to one another of two lines* puts angle in the category of *relation*, says Proclus, and there are objections to this.

We overcome the difficulty by defining an angle as that which is formed by three ordered points A, O, B, where A and O are distinct, and B and O are distinct. We call O the vertex of the angle, and then adopt an axiom which says that the angle can be measured.

Euclid's Fourth Postulate, that *all right angles are equal to one another*, says Proclus, states an intrinsic property of right angles, and is therefore either an axiom, according to Geminus, or a theorem, according to Aristotle. Proclus gives a long discussion of the difference between axioms, definitions and postulates, but we must omit this. It is interesting to note, however, that Hilbert, in his work on the foundations of Euclidean geometry, *proves* that all right angles are equal, thus confirming Aristotle's opinion.

We have already mentioned the attention given by Proclus to Postulate 5 (p. 159). Proclus devotes a lot of space to an examination of the Postulate, believing, with Ptolemy, whom he found to be wrong, that it can be proved.

The role of the infinite and the concept of infinite divisibility in mathematical science are discussed at length, and Proclus is constantly introducing his audience to differences of opinion among recognized authors, to proofs alternative to those given by Euclid, and to criticisms of his selection or ordering of proofs.

Proclus also examines questions still basic to what we now call *the philosophy of mathematics*, the nature of the objects of mathematical inquiry, and the character and the validity of the procedures used by mathematicians in handling these objects. Proclus' commentary, says the author of the excellent translation we have been using, is the only systematic treatise surviving from antiquity which deals with these questions.

Let us now try to communicate the flavour of parts of the Proclus commentary. In discussing *a point is that which has no part*, Proclus says:*

*Morrow, p. 75.

We have expanded largely on these matters to show that points, and limits in general, have power in the cosmos, and that they have the premier rank in the *All* by virtue of carrying the likenesses of the first and most sovereign cause . . .

Discussing the definition of a circle, Proclus says:

The first and simplest and most perfect of the figures is the circle . . . If you divide the universe into the heavens and the world of generation, you will assign the circular form to the heavens and the straight line to the world of generation . . .

Later on,

Pythagoreans assert that the triangle is the ultimate source of generation and of the production of species amongst things generated . . . The Pythagoreans thought that the square more than any other four-sided figure carries the image of the divine nature. It is their favourite figure, indicating immaculate worth, for the rightness of the angles imitates integrity, and the quality of the sides abiding power.*

We have different feelings about squares nowadays, but one thing is plain: we have lost our sense of wonder concerning the nature of simple geometric figures, and this is a definite loss. But neo-Pythagorean doctrines seem to be developing in the subcultures of young people nowadays.

Exercises

1. Assuming the Postulate that a unique line can be drawn through two distinct points P and Q, prove that two distinct lines intersect in at most one point. (Use *reductio ad absurdum.*)

2. Show that two triangles ABC, $A'B'C'$ are not necessarily congruent if $AB = A'B'$, $AC = A'C'$ and angle ABC = angle $A'B'C'$. (This shows that in the S-A-S theorem, the angle must be *included* between the given sides. Set the point of the compasses at A, extend them

*Morrow, pp. 117, 131 and 136.

to C, and observe that unless angle ACB is a right angle, there is another point $C' \neq C$ on BC such that $AC = AC'$.)

3. The lines ABC and DEF are such that the angle ABE is equal to the angle BEF. Prove that the lines ABC and DEF are parallel. (This is Euclid, Proposition 27, that if a straight line falling on two straight lines makes the alternate angles equal, then the two straight lines are parallel. Assume that the lines do meet, and then assume also the theorem that the external angle of a triangle exceeds either interior and opposite angle. A contradiction is therefore reached. Therefore . . .)

4. If the lines ABC and DEF are parallel, prove that the angles ABE and BEF are equal. (This is Euclid, Proposition 29. Since the angle on a straight line is two right angles, an alternative statement is that the angles CBE and BEF add to two right angles. Assume that this is *not* so, and now invoke Postulate 5. If the sum is less than two right angles, the lines, when produced, intersect on one side of BE, and if the sum exceeds two right angles, the lines, when produced, intersect on the other side of BE. A contradiction! Therefore . . .)

5. Assuming Postulate 5, in the form of the Playfair Axiom, prove that the sum of the angles of a triangle ABC is two right angles. (Through B draw the unique parallel BD to the line CA, and use the theorems already proved for angles made by a transversal with a pair of parallel lines.)

6. Show that the theorem of Exercise 5 is equivalent to the theorem that the exterior angle at B of the triangle ABC is equal to the sum of the two interior and opposite angles BCA and CAB. (Compare this theorem with the one, mentioned in Exercise 3, of *absolute* geometry.)

7. To show that the Playfair Axiom is equivalent to Euclid's Postulate 5, you must show that if the Playfair Axiom is valid, then so is Postulate 5, and conversely, if Postulate 5 is valid, so is the Playfair Axiom. See how far you can proceed with either proof.

8. If P is a point on a circle, centre O, and PQ is a line through P which does not intersect the circle again (this is called the *tangent* to the circle at P), prove that the angle OPQ must be a right angle. (Show that if the angle OPQ is *not* a right angle, then there is a point $R \neq P$ on PQ such that angle ORP = angle OPQ, and therefore R also lies on the circle. Contradiction!)

9. A, B, C are points on a circle, centre O. Join A and B to O, and also C to O. Mark the equal angles in the isosceles triangle $ACO (OC = OA)$, and the isosceles triangle $BCO (OC = OB)$, and use Exercise 6 to show that angle $AOB = 2$(angle ACB), so that the angle ACB is constant as C moves on each of the two arcs terminated by A and B.

If C and C' are points on each arc, deduce that the sum of angles ACB and $AC'B$ is two right angles.

·7·

CARTESIAN AND PROJECTIVE GEOMETRY

ALTHOUGH we have not dwelt on it, Greek geometry, powerful although it undoubtedly was, had to consider every possible special case of a theorem before the proof was recognized as complete. An enormous simplification came into being with the introduction of *sensed* magnitudes, which could be either positive or negative, and the introduction of *co-ordinates* by René Descartes in 1637 changed the face of geometry for all time. What Descartes does is to fix the position of a point in a plane by its distances from two fixed lines, drawn at right angles to each other through a point O. The lines are called the *axes* of co-ordinates, and O is called the *origin* of co-ordinates. We say that a point P has the co-ordinates (x, y) if the distance of P from the OY-axis is x, and the distance of P from the OX-axis is y (Figure 67). The numbers x are taken to be *negative* if P is in the second quadrant or third quadrant, and the numbers y are taken as negative if the point P lies in the third or fourth quadrant. The diagram shows the four possible combinations of sign corresponding to the four quadrants of the plane.

If P moves on a curve, there is a *relation* between x and y, and this is called the *equation* of the curve. For instance, if P moves on the line shown, the relation is $x+y-1 = 0$, and we say that the equation of the line is $X+Y-1 = 0$. A point (x, y) lies on this line if and only if *the co-ordinates satisfy the equation of the curve*, in this case if and only if $x+y-1 = 0$.

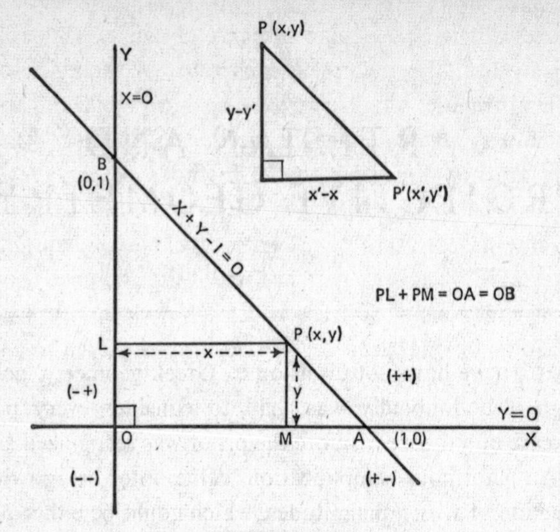

Figure 67

The distance between two points $P = (x, y)$ and $P' = (x', y')$ is given, using the Pythagoras Theorem, by the formula

$$(PP')^2 = (x-x')^2+(y-y')^2,$$

and any book on *analytical* geometry, as it is called, shows what a wealth of theorems can be proved by the methods initiated by Descartes, as a sideline, more or less, to his real interest, which was philosophy. Of course Descartes did use some theorems of Euclid in obtaining his wide-ranging results, and in a letter to the Princess Elizabeth of Bohemia, with whom he corresponded on both mathematical and philosophical matters, Descartes says:

... I use no theorems except those which assert that the sides of similar triangles are proportional, and that in a right triangle the square of the hypotenuse is equal to the sum of the squares of the sides. I do not hesitate to introduce several unknown quantities, so as to reduce the question to such terms that it shall depend only on these two theorems.

Descartes discusses the conic sections, of course, and, since he is using algebra, the theory of equations. Without Descartes, modern mathematics would not have been possible. But there is still controversy as to whether the methods of pure (or *synthetic*) geometry are not preferable to the methods of *analytical* geometry. Geometers are concerned with the *attractiveness* of a proof. Some proofs using co-ordinates are long and turgid, and can be replaced by a short synthetic proof. But the true geometer tries to be conscious of both methods of approach to a problem, and a solution by one method often illuminates the other.

We shall not do more here than give the equations of some curves we have encountered, the circle and the three types of conic section. A circle centre at the origin (0, 0) and radius r has the equation:

$$X^2+Y^2-r^2 = 0.$$

If the centre is at the point (p, q), the equation is

$$(X-p)^2+(Y-q)^2-r^2 = 0.$$

The equations of curves depend very much on where the axes of co-ordinates are chosen. For suitable axes, the equation of an ellipse is:

$$X^2/a^2+Y^2/b^2 = 1,$$

that of an hyperbola is

$$X^2/a^2-Y^2/b^2 = 1,$$

and a parabola is given by the equation

$$Y^2 = aX.$$

The respective axes are shown in Figure 68. These equations mask the focal properties of the conic sections, which we discuss in the next chapter, and often students, whose only introduction to the conic sections is via analytical geometry, never discover the

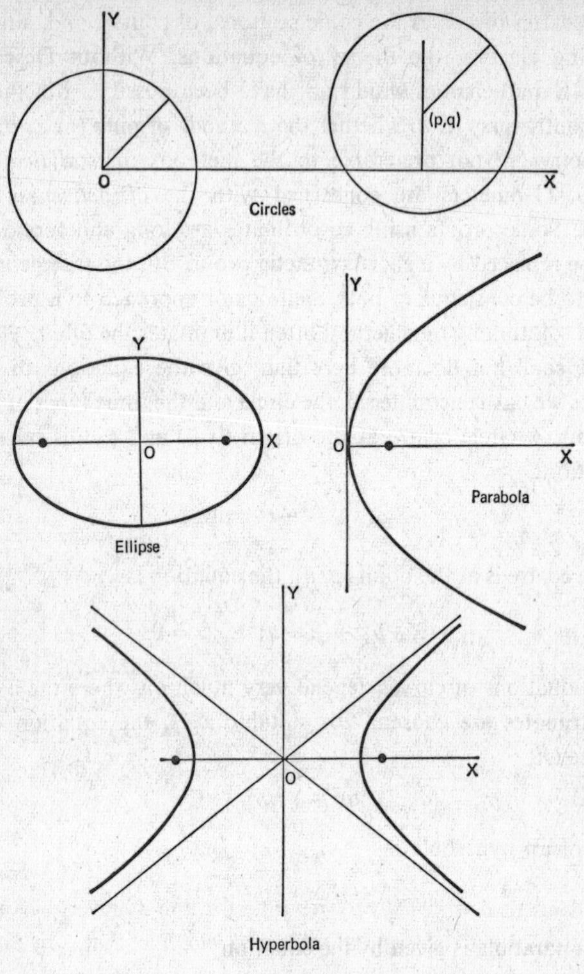

Figure 68

fascinating properties hidden by their simple equations. A synthetic approach, using pure geometry, is advantageous in this case.

The geometry of three dimensions can be studied by the methods of Descartes, and Figure 69 shows how co-ordinates (x, y, z) can

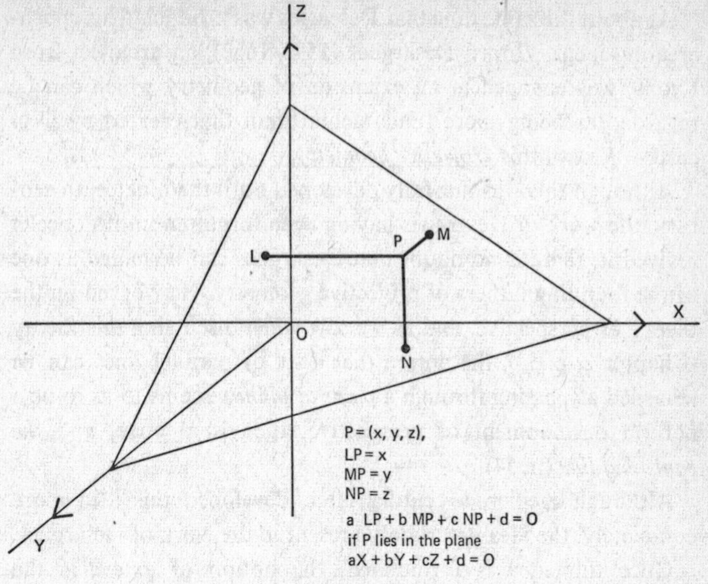

Figure 69

be attached to any point in space of three dimensions. Again, the use of positive and negative numbers enables us to discuss the eight octants into which three mutually perpendicular planes drawn through a point O divide space. In space of three dimensions the equation of a *plane* is of the form

$$aX+bY+cZ+d = 0,$$

and we deal with *lines* by regarding a line as given by the intersection of two distinct planes through the line, that is by the simultaneous equations

$$aX+bY+cZ+d = 0 = a'X+b'Y+c'Z+d'.$$

Again, an enormous field of human endeavour opens before us, which we cannot enter here. We venture into higher dimensions in Chapter 9.

At about the same time that Descartes was producing his epoch-creating ideas, Girard Desargues (1593–1662), an architect from Lyons, was engaged in an extension of geometry which can be regarded as being more fundamental than that created by Descartes. We call this *projective geometry.*

Although this was not fully developed until the nineteenth century, the work of Desargues having been forgotten until Poncelet revived it, there is no doubt that Desargues can be hailed as one of the founding fathers of projective geometry. He worked on the theory of perspective, and as we saw when discussing this theory (Chapter 2, p. 52), the notion that a set of parallel lines can be regarded as passing through a *point at infinity* seems to us to be a natural development of perspective, the visual cone, and *the vanishing line* (p. 54).

Although Kepler, a century later, developed this idea more completely, the idea is certainly present in the work of Desargues.

Once this idea is formulated, the notion of *extending* the Euclidean plane by adding to it *points at infinity* is bound to arise, and these points are considered to lie on a line, the *line at infinity.*

If we then democratize the extended plane, and consider *all* points to have the same status, we are radically changing the nature of our plane. Points at infinity, which are sometimes called *ideal* points, become ordinary points, and the *ideal* line, the line at infinity, becomes an ordinary, common or garden line. The behaviour of points and lines in this extended plane now satisfies the two theorems, or axioms, depending on our point of view:

1. Two distinct points in the plane determine a unique line, on which they both lie,
2. Two distinct lines in the plane determine a unique point, through which they both pass.

There is no longer any reference to the possibility of lines being parallel. This geometry of the extended plane is one of points, lines and intersections. Any theorem on points, lines and intersections which is true in our plane remains true if we project the figure from a vertex onto a picture plane, as in Figure 12. Hence the title *projective geometry*.

Now, if it is asserted that this geometry, in which there is no reference to measurement, comprises the whole of Euclidean geometry, the reader may well be startled, and murmur: 'But are there *any* theorems which do not involve measurement, but only points, lines and intersections?' Our discussion of Euclid in Chapter 6 did not reveal any, but there *are* some theorems which can be found in Euclid which belong to projective geometry. None is as evidently a theorem of projective geometry as the following one, not in Euclid, but assigned to Pappus of Alexandria, who lived several hundreds of years later.

Two straight lines l and m lie in a plane, and the distinct points A, B, C lie on l, while the distinct points A', B' and C' lie on m. Then the three intersections of the pairs of lines AB', $A'B$; BC', $B'C$; and CA', $C'A$ lie on a line (Figure 70).

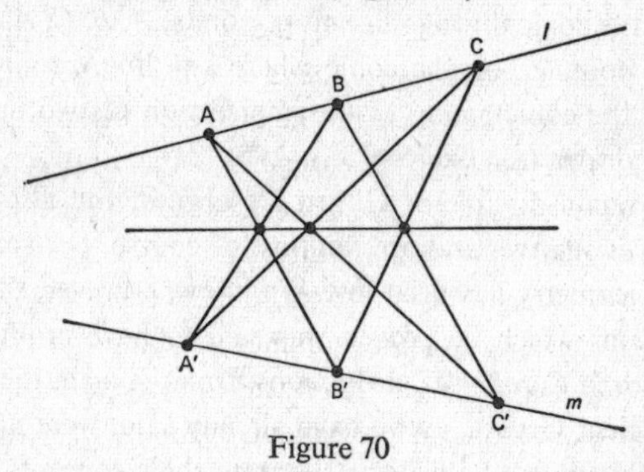

Figure 70

This theorem is of fundamental importance nowadays in any discussion of the foundations of geometry. Over a thousand years later Blaise Pascal, at the age of sixteen, formulated a theorem, also projective, which can be regarded as a generalization of the Pappus Theorem: the six distinct points A, B, C, A', B' and C' lie on a conic. Then the three intersections of the pairs of lines AB', $A'B$; BC', $B'C$; and CA', $C'A$ lie on a line (Figure 71). Note that this

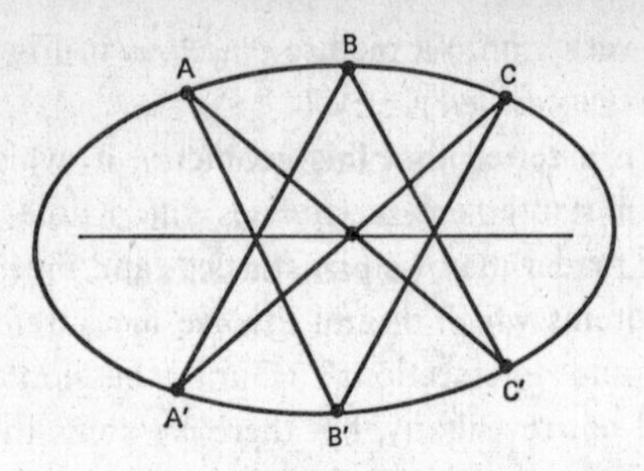

Figure 71

theorem is true for *any* conic, for a circle, for an ellipse, for a parabola and for a hyperbola, which can all be produced as sections of a circular cone.

Also, with a reasonable definition of *tangent* at a point of a conic, the theorem continues to hold when some of the six points are not distinct from each other. The theorem can be used to *construct* the conic through five given points, A, B, C, A', B', for it gives the point C' on the conic where any line through B, say, intersects the conic again, as the intersection of two lines which are easily drawn (see Exercise 3, p. 195).

Pascal was a disciple of Girard Desargues, and also a writer. His *Pensées* survive, and are famous, of course, but some of his work on geometry has been lost. We know, however, that one of the theorems which he proved was said to have produced over three hundred *Corollaries*, deductions from the main theorem.

The ancient Greeks, as we have already said, were not able to produce theorems of this type, because their concepts were not sufficiently advanced, and every special case had to be examined afresh. At its peak, with the work of Apollonius, Greek geometry was a marvellous structure, with a strength still underrated by those who think that everything modern is bound to be better, even if used by incompetents. But *general* theorems are lacking in Greek geometry.

Desargues has another theorem of projective geometry associated with his name, and this is also one of fundamental importance in the foundations of geometry. ABC, $A'B'C'$ are two triangles, not necessarily in the same plane, which are such that the joins AA', BB' and CC' all pass through the same point V. Then the intersections of the three pairs of sides BC, $B'C'$; CA, $C'A'$; and AB, $A'B'$ are three points on a line (Figure 72).

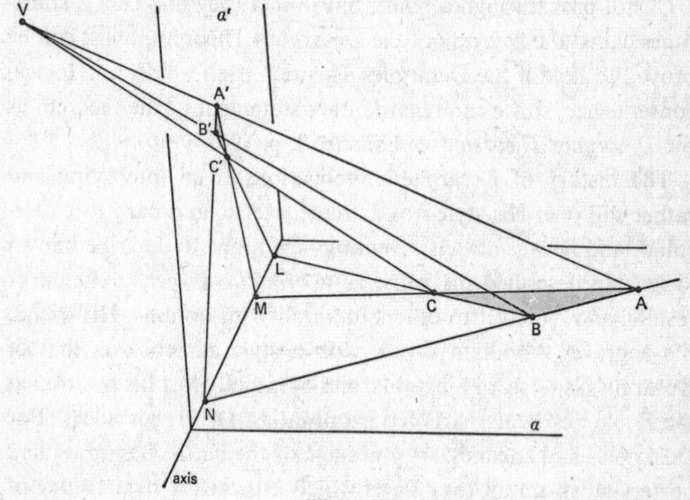

Figure 72

To explain how this theorem might well be observed by anyone working on the theory of perspective, as Desargues did, we note that if ABC is a triangle in the plane α, and the image of ABC in the plane α' is the triangle $A'B'C'$, and V is the vertex of the visual cone (the eye, in fact), then AA', BB' and CC' all pass through V, and since AB projects into $A'B'$, and points on the axis of projection *project into themselves*, the intersection of AB and $A'B'$ is *a point on the axis of projection.*

This is true for the intersection of BC, $B'C'$ and for the intersection of CA, $C'A'$. Hence the three points designated in the statement of the theorem are indeed on a line.

Now, what happens if we keep triangle ABC fixed in the plane

α, and $A'B'C'$ fixed in the plane α', and *rotate the plane* α about the axis of intersection of α and α' until α *coincides* with the plane α'?

The two triangles ABC, $A'B'C'$ now lie in a plane, and they are still such that the intersections of BC, $B'C'$; CA, $C'A'$; and AB, $A'B'$ lie on a line, the former axis of intersection of α and α'. We are not immediately convinced that the joins AA', BB' and CC' still pass through a point, but in fact they do. This is sometimes called the *converse* of the Desargues Theorem, but it can be proved to hold if the Desargues Theorem itself holds, as a logical consequence, so we can regard either statement of the theorem as the *Desargues Theorem* (see Exercise 2, p. 194).

The history of Desargues' publications is an interesting and rather sad one. His style was extremely terse, and hard to understand, and he was always inventing new terms to describe known ideas. He described his work as a *brouillon-projet*, a sketch of results, and he left it to others to amplify his notions. His friend, the engraver Abraham Bosse, with a style as prolix as that of Desargues is terse, did just this, and eventually lost his position at the École des Beaux Arts for propagating Desargues' ideas. For Desargues had enemies. At one stage of the battle Desargues had posters stuck up all over Paris which attacked a rival theory of perspective attributed to a *père* Dubreuil, and the posters offered a reward to anyone who could prove that the Desargues theory was not the best! Mathematicians have often engaged in rancorous controversy, but Girard Desargues seems to have been unique in his methods of disputation.

Although praised by Descartes, Pascal and others, the work of Desargues vanished for centuries until an army engineer, by that time *le général* Poncelet, revived it in 1822, in his own great treatise on projective geometry, which was mostly written while Poncelet was a prisoner of the Russians, captured during Napoleon's disastrous retreat from Moscow.

In the second edition of the *Traité des Propriétés Projectives des Figures*, Poncelet joins the ranks of the oppressed. Page after page is devoted to controversy, the objects of the attack being not only certain Academies, but two other French geometers, Gergonne and Brianchon. The subject disputed this time is not perspective, but the *Principle of Duality*, one of the great notions of projective geometry.

Before we describe this Principle, it may be remarked that the assertion in a recent *New Yorker* article* that mathematicians are never dismayed if someone else first publishes a cherished discovery, or lays claim to it, must be based on a complete ignorance of the history of mathematics!

The Principle of Duality is based on the remark that our two main axioms for projective geometry of the plane, 1 and 2 on p. 178, show a *duality*, that is they can be converted, the one into the other, if we merely interchange the terms *point* and *line*, and the words *lie* and *pass*, and, of course, the words *on* and *through*. If we use the general term *incident* for a point which lies on a line, or a line which passes through a point, and rewrite the axioms in the form:

1. Two distinct points in the plane determine a unique line with which they are both incident,
2. Two distinct lines in the plane determine a unique point, with which they are both incident,

then the two axioms are inter-convertible even more simply.

Now, suppose that we add to our simple dictionary of interchangeable terms *join* and *intersection*, *triangle*, referring to three non-collinear points, and *triline* referring to three non-concurrent lines, and of course *collinear* referring to points on a line, and *concurrent* referring to lines through a point, then we are in a position to apply the Principle of Duality to theorems which involve only points, lines, intersections, joins, triangles and trilines.

The Principle asserts that every such theorem has a shadow (dual) theorem involved with itself, a kind of mathematical *Doppel-Gaenger*, and that this dual theorem arises from the mere

*19 February 1972, p. 39.

enunciation of the original theorem by *an automatic interchange*, with the help of our little dictionary, of the terms *point*, *line*, or *join*, *intersection*, or *collinear*, *concurrent*, or *triangle*, *triline*.

When this Principle is applied to the Desargues Theorem, we obtain the *converse* to the Desargues Theorem. Let us verify this by rewriting the Desargues Theorem in the form:

If two triangles are such that the joins of corresponding vertices are incident with a point, then the intersections of corresponding sides are incident with a line.

On transformation, using our dictionary, and interchanging *vertex* and *side*, we have:

If two trilines are such that the intersections of corresponding sides are incident with a line, then the joins of corresponding vertices are incident with a point.

This is the converse of the Desargues Theorem. Mathematics tries to do without principles as it advances, and the Principle of Duality is a theorem nowadays, perhaps most easily proved when algebra is introduced into geometry, but a discussion of this would take us too far. However, even this short description of the Principle of Duality, which can be extended to projective space of any number of dimensions, may show the eagerness of geometers involved with its invention (or discovery) to lay claim to it as their very own idea.

We shall now show the importance of the Desargues Theorem in the theory of perspective. Once again, as on p. 181 we suppose that we have a triangle ABC in a plane α, and that the triangle $A'B'C'$ is its representation in a plane α', the eye of the painter being V. Corresponding lines, such as AB and $A'B'$, intersect on the axis of projection, the intersection of the planes α and α'. We again fold the planes α and α' about the axis, until they lie in the same plane. This process is called *rabattment*.

We now have two triangles ABC, $A'B'C'$ in the same plane, and the intersections of AB, $A'B'$, of BC, $B'C'$ and of CA, $C'A'$

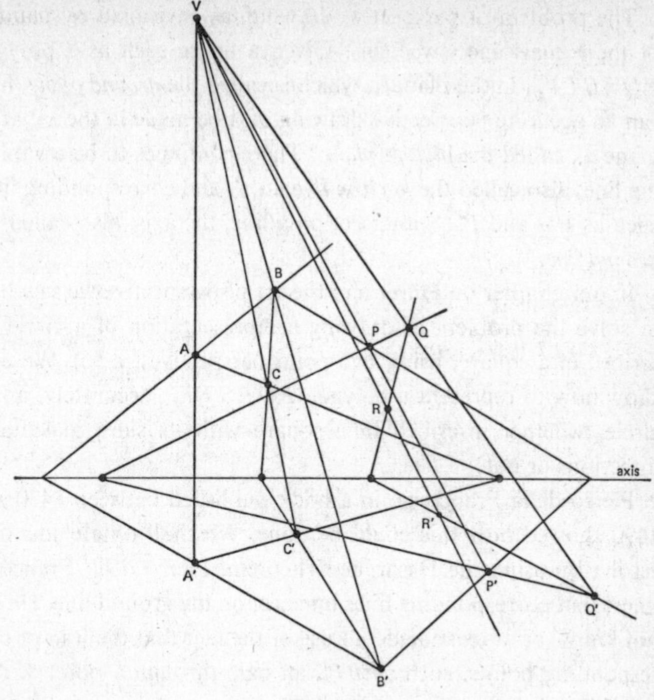

Figure 73

lie on a line. By the Desargues Theorem, the joins AA', BB', CC' again pass through a point. We once more call this point V (Figure 73). Now V is uniquely determined as the intersection, say, of the lines AA' and BB'. It follows that if we have any polygon $PQRSTU\ldots$ and the projected image $P'Q'R'S'T'U'\ldots$, then corresponding lines such as PQ, $P'Q'$ all meet on the axis, and the joins of *all* pairs of corresponding points, PP', QQ', RR', SS', ..., all pass through the point V. This last result is an immediate consequence of applying the Desargues Theorem to the triangles ABP, $A'B'P'$, then to the triangles ABQ, $A'B'Q'$, and so on. The joins PP' and QQ' will pass through the intersection of AA' and BB', which is the point V.

The problem of perspective drawing, as envisaged by painters of the Renaissance, was this. Given a figure such as a polygon $PQRST\ldots$ in the plane α, which is called the *ground plane*, how can an accurate perspective drawing of it be made in the rabatted plane α', called the *picture plane*? There continues to be a vanishing line, also called the *horizon line*, in α', and corresponding lines such as PQ and $P'Q'$ intersect on a line, the axis, also called the *ground line*.

In our chapter on Dürer and the art of perspective we saw how to solve the problem of drawing a representation of a circle inscribed in a square, using two-point perspective (p. 56). We now show how to represent a polygon $PQRST\ldots$ accurately, and a circle, whether inscribed in a square with its sides in suitable directions or not.

Pierro della Francesca, in a book published between 1470 and 1490 showed how this could be done. We shall obtain identical results by using the Desargues Theorem. Pierro della Francesca knew that corresponding lines intersect on the ground line. He did not know, or at least made no use of the fact that the join of corresponding points, such as PP', all pass through a point V. We call this point the *centre of perspective*.

The one additional theorem we shall use is based on the idea that our perspective drawing is *a map* of the given drawing which lies in the same plane. So we can say that points on the vanishing line are mapped onto points *at infinity*. Suppose that a line PQ intersects the vanishing line v at a point T (Figure 74). The points of the line PQ are mapped onto the points of a line $P'Q'$, and T, which is on PQ, is mapped onto a point T' *on* $P'Q'$, but *at infinity* on this line. Now the join TT' passes through V, since T and T' are corresponding points. Hence *the line* VT *intersects* P′Q′ *at infinity*. In other words, *the mapped line* P′Q′ *is parallel to the line* VT.

We are now in a position to make a perspective drawing of any given figure in the ground plane, after choosing *either* the axis *or*

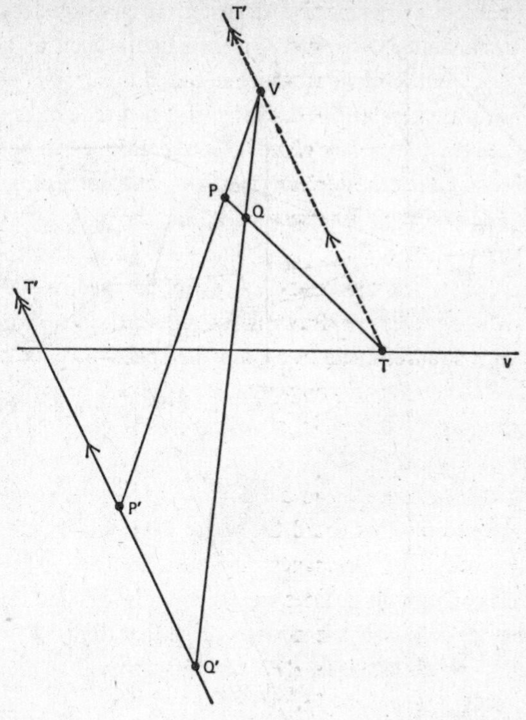

Figure 74

the vanishing line, which is parallel to the axis, and the centre of perspective V.

We reproduce a figure from the work on perspective by Pierro della Francesca, and show how our methods give the perspective drawing of two sides of the regular pentagon mapped by Pierro (Figure 75). To avoid a prolixity of lines, we do not give the whole pentagon and its map, but it can be seen that we obtain the same results.

Pierro takes a square $BCED$ and a regular pentagon $FGHIK$ inside the square, and shows how to make a perspective drawing of this figure. He takes the axis as the line BC, so that $B' = B$, and

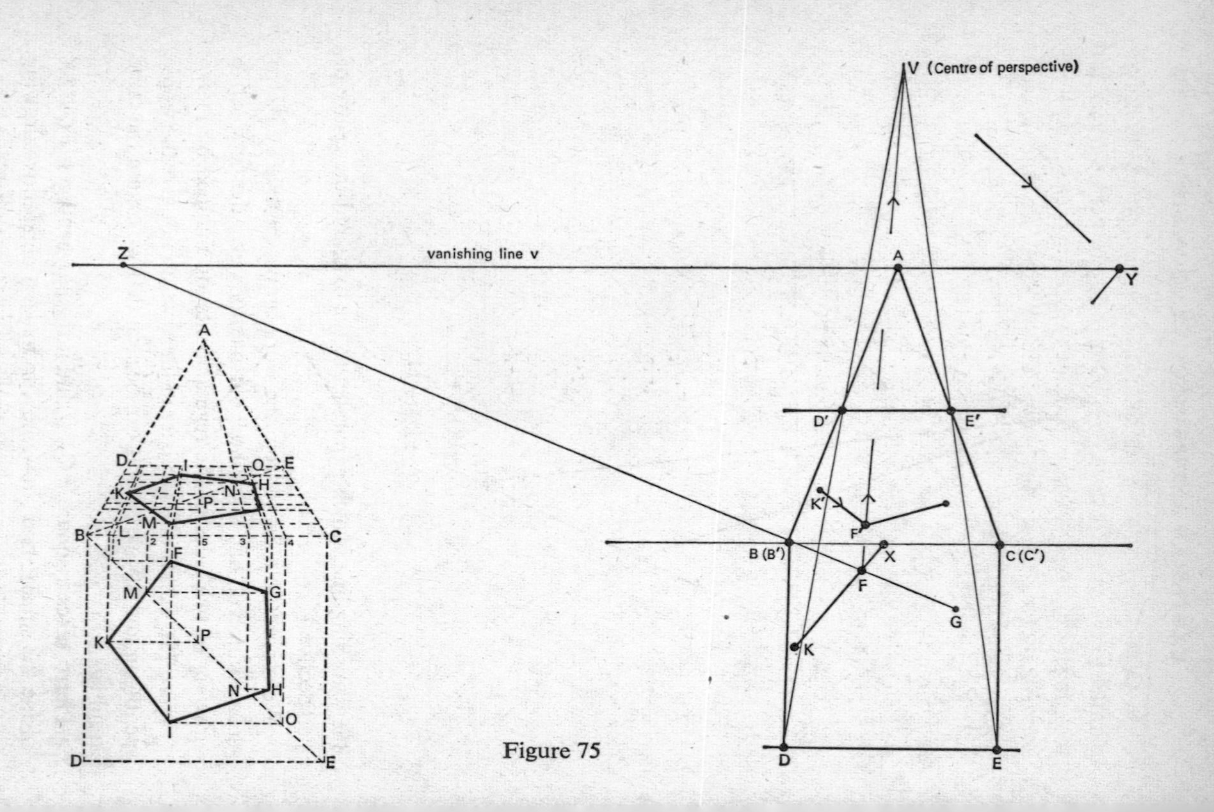
V (Centre of perspective)
vanishing line v
Z
Y
A
D′
E′
K′
F′
B (B′)
C (C′)
X
F
G
K
D
E
A
D
I
Q
E
K
N
H
P
M
B
L
C
1
2
5
3
4
F
M
G
K
P
N
H
O
D
E

Figure 75

$C' = C$, and his point A is on the vanishing line, which we draw parallel to the axis BC through A. Our centre of perspective V is the intersection of DD' and EE', and we can construct the perspective drawing using either V and the axis, or V and the vanishing line v, or both. We make use of both.

If KF meets the axis in the point X, then $K'F'$ passes through X. If KF meets the vanishing line v in Y, then the line $K'F'$ is parallel to VY. So, to obtain K', which lies on VK, we find the intersection of VK and the line through X parallel to VY. To find F', we obtain the intersection of VF and the line through X which is parallel to VY, and similarly for the points G', H' and I'. This gives a rapid method for constructing plane perspective drawings.

Pierro della Francesca also showed how to represent three-dimensional surfaces, such as cubes, in a perspective drawing, but we shall not go any further into this than we did in Chapter 2. On the other hand, a perspective drawing of a circle is of great interest because of the different curves we obtain, the conic sec-

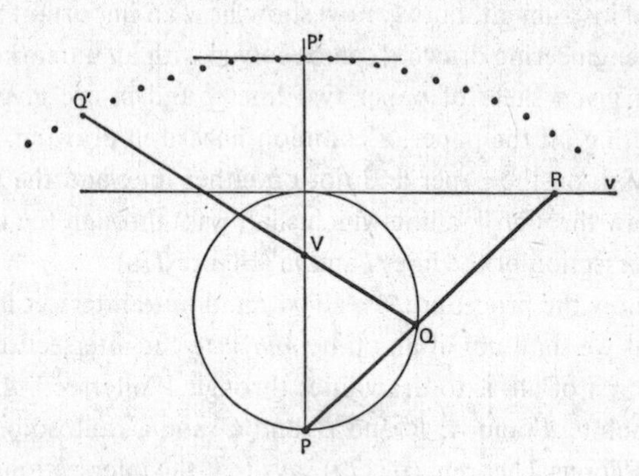

Figure 76

tions in fact, depending on the relation of the circle to the vanishing line. In Chapter 2, p. 56, the circle inscribed in a square becomes an ellipse. Here we show the *parabola* obtained when the circle we are representing *touches* the vanishing line (Figure 76).

We *choose* V, in this case, on an axis of symmetry for the circle and the vanishing line, and *choose* P', which is on VP, where P is the other end of the diameter through V not on v, to lie on the paper. If Q is now any point on the circle, and PQ meets v at R, then we know that VR is parallel to $P'Q'$, and that Q' lies on VQ. We are thus able to construct any number of points on the parabola which is the perspective image of the circle.

Choosing the map P' of one point on the circle enabled us to dispense with the axis, but it can be verified that all pairs of corresponding lines in the figure meet on a line.

We can, of course, carry out the construction using only V and the axis, and then all mapped points arise as the intersection of pairs of lines, for which we may use an unmarked ruler. The Desargues Theorem, as we noted, was stressed by Poncelet in his great treatise on projective geometry, and projective geometry is often called *the geometry of the ungraduated ruler*. It might be thought that not much can be done with such a primitive geometrical instrument, but we now show how an important problem in civil engineering drawing can be solved with an unmarked ruler.

On a given sheet of paper two lines l and m are given, these intersecting off the paper, a common hazard in drawing. A point V is given, on the paper and not on either line, and the problem is to draw through V a line which shall pass through the inaccessible intersection of the lines l and m (Figure 77a).

To show the procedure, we allow l and m to intersect in Figure 77b, but we shall not at any time *join* V to the intersection. What we do, first of all, is to draw lines through V intersecting l and m in the points A' and A, B' and B, and C' and C, and so on.

The Pappus Theorem (p. 179) says that the intersections of $A'B$ and AB', of $B'C$ and BC' and of $C'A$ and CA' lie on a line. Since a line is uniquely determined by two points, we obtain points on the *same* line if we find the intersection of, say, $C'D$ and CD', so that as we draw lines through V cutting l and m, the intersection of cross-joins always lies on a line.

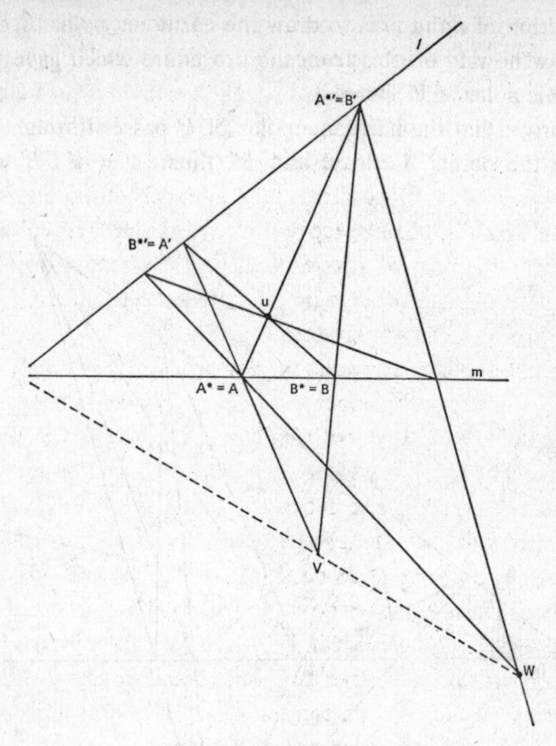

Figure 77a

Now suppose that our D and D' both coincide with the intersection of l and m. Then $C'D$ is l, and CD' is m, and the intersection of these two lines, l and m, is a point on the line obtained from our cross-joins. To give this line a name, we call it *the harmonic polar of* V *with respect to* l *and* m.

Now let U be the intersection of $A'B$ and AB', so that U is a point on the harmonic polar of V. Then U has a harmonic polar with respect to l and m. We show that *this line passes through* V. In other words, if the harmonic polar of V passes through U, then the harmonic polar of U passes through V. When we have proved this, all that we need to do to obtain the line through V and the intersection of l and m is to draw the harmonic polar of U, and we know how to do this from the procedure which gave us the harmonic polar of V.

To prove that the harmonic polar of U passes through V, we *rename* the points A, A', B and B', thus: $A = A^*$, $B' = A^{*\prime}$,

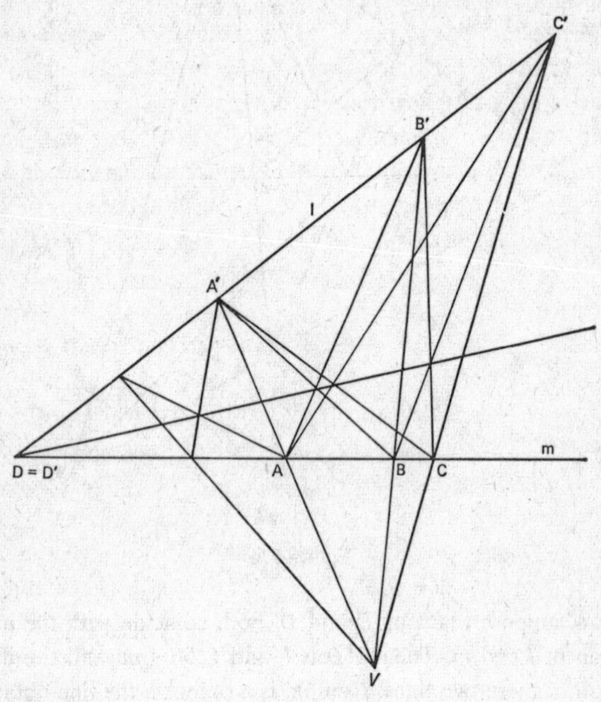

Figure 77b

$B = B^*$ and $A' = B^{*\prime}$. Then, by our method, the harmonic polar of U contains the intersection of the lines $A^* B^{*\prime}$ and $A^{*\prime} B^*$, that is, of the lines AA' and BB', and this is the point V.

Another method of construction, using the converse of the Desargues Theorem, is indicated in the Exercises.

The importance of projective and Euclidean geometry in civil engineering has always been manifest, and many of the great French geometers were attached to the army. In fact, even Napoleon Buonaparte has some theorems named after him. Although Girard Desargues was not an army officer, he worked on the fortifications of La Rochelle, and Descartes visited him and inspected the fortifications in 1628.

The statement we made earlier, that projective geometry includes Euclidean geometry cannot be proved here, since we should have to introduce new concepts such as complex numbers, cross-ratio and so on. It was Poncelet who showed that *all circles pass through the same pair of complex points on the line at infinity.* These are usually called *I* and *J*, and, with all due reverence, the full names are understood to be Isaac and Jacob. Then Laguerre, while still a schoolboy, showed that Euclidean angle can be defined with regard to *I* and *J*, and the way was clear to proving that *Euclidean geometry is merely projective geometry with reference to a special pair of points.*

Finally, as an indication of the directions in which geometry has developed, we point out that it is possible to attach the label *points* to an otherwise unspecified set of objects, and the label *lines* to certain subsets of this *universal* set, the name suggesting that we shall not consider anything outside this set. We say that a point *P* is *incident* with a line *l*, if the point *P* lies in the subset which specifies the line *l*.

Now if these points and lines satisfy certain axioms, we have a structure which we call *a geometry*, and we can apply our deductions to any set of objects which, with suitable interpretations of the terms *point*, *line* and *incident* satisfy the axioms. This is a far cry from Euclid.

Consider the following problem. A tennis match is played between two teams, each player playing one or more members of the other team. We are told that:

1. any two members of the same team have exactly one opponent in common,

2. no two members of the same team play all the members of the other team between them.

Prove that two players who do not play each other have the same number of opponents. Deduce that any two players, whether in the same or different teams, have the same number of opponents.

This problem can be considered as a geometrical problem if we label the members of one team *points*, and the members of the other team *lines*. We then say that a point is *incident* with a line if the two players specified *play each other*. This is an extension of our generalization above, where all the objects considered are in the same set.

The condition 1. is now seen to imply our two fundamental axioms of projective geometry of the plane (p. 183), two distinct points determine a unique line with which they are both incident, and two distinct lines determine a unique point with which they are both incident. The condition 2. for the tennis match leads to the existence of four points, at least, no three of which are on a line! From this additional property of the points and lines, and the two fundamental axioms, the points and lines satisfy the requirements of what we call an *abstract projective geometry*, and the problem can be solved.

The reasoning involved is not simple, and, above all, no information is to be gathered from diagrams which *assume* results which have to be proved, or may be incorrect. The reader may be able to solve this problem by other methods.

Exercises

1. The Pappus Theorem (p. 179) considers sets of three distinct points on two distinct lines. The dual theorem deals with sets of three distinct lines through two distinct points. Formulate the dual theorem, and verify it by drawing. (If you are ingenious, you will find two lines in your dual figure, each containing three distinct points, and an application of the Pappus Theorem will prove your dual theorem.)

2. Assume that you wish to prove the converse of the Desargues

Theorem, and that in Figure 72 the points L, M and N are on a line, but we wish to prove that the lines AA', BB' and CC' pass through one point. Apply the Desargues Theorem to the triangles AMA' and BLB'. (The joins AB, ML and $A'B'$ of corresponding vertices pass through the point N, so the theorem is applicable.) This should prove that AA', BB' and CC' all pass through one point.

3. Take five points A, B, C, A' and B', and use Pascal's Theorem to determine where a line drawn through B meets the conic through the five points again. (The theorem is used to find C'. By varying the line through B, any number of points may be constructed on the conic, using only an unmarked ruler.)

4. Solve the construction of p. 190, to draw a line through V and a non-accessible intersection, by the use of the Desargues Theorem. (Draw a triangle VAB, with the vertex A on l, and the vertex B on m, and let LMN be any suitable line cutting the side AB in L, the side VB in M, and the side VA in N. Now draw a triangle $V'A'B'$ with A' on l, $A'V'$ passing through N, with B' on m, and $A'L$ passing through B' (this determines B'), and finally $B'V'$ has to pass through M, which determines V' as the intersection of $A'N$ and $B'M$. The triangles VAB and $V'A'B'$ now satisfy the requirements for the converse of the Desargues Theorem, and so VV', AA' and BB' pass through a point. But this point is the intersection of l and m.)

5. In the diagram of the Desargues Theorem (Figure 72), there are ten points and ten lines. Every line contains three points, and three lines pass through every point. The triangles ABC, $A'B'C'$ are *in perspective* from V, as *centre* of perspective, and LMN is the *axis* of perspective. Show that *any* one of the ten points of the *configuration* (as it is called) may be chosen as centre of perspective, and then two triangles can be found among the remaining nine points which are in perspective from this point, and their axis of perspective also consists of three points (collinear, of course) of the configuration. For example, if N is chosen as centre of perspective, the perspective triangles are BLB' and AMA', and the axis is $C'VC$. Reproduce the configuration ten times on a large sheet of paper, and colour the two perspective triangles corresponding to a different choice of centre of perspective in each case. You will find the total pattern an attractive one.

·8·

CURVES

CURVES have a fascinating history, and this chapter will be devoted to some of them, beginning with circles, and proceeding through the conic sections to a necessarily brief survey of a few of the delightful constructs of the human mind in the past two thousand years. The straight line and the circle, or near approximations to these, appear in nature, and so does the spiral, as we shall see (p. 251). Once spheres were produced, conic sections could be observed as shadows, and so it was natural for these curves to be studied. But the ancient Greeks invented other curves as well.

We begin with the circle, which necessarily comes first, since its mathematical definition is so simple. We give some properties of the circle, since the methods developed will be of use with other curves.

If O is the centre of a given circle, and P a point on it, the line through P at right angles to OP does not intersect the circle again, and is called the *tangent* to the circle at P (Figure 78). To prove this theorem, we assume that in fact this line does intersect the circle at a point $Q \neq P$. Since $OP = OQ$, the angles OPQ and OQP are equal, and since angle OPQ is a right angle, by hypothesis, so is the angle OQP. But the angles of triangle OPQ add up to two right angles. It follows that angle POQ is zero, and so $Q \neq P$ has to be rejected. (In Exercise 8, p. 171 the reader was asked to prove the converse of this theorem.)

The circle is the unique curve for which the tangent at *every*

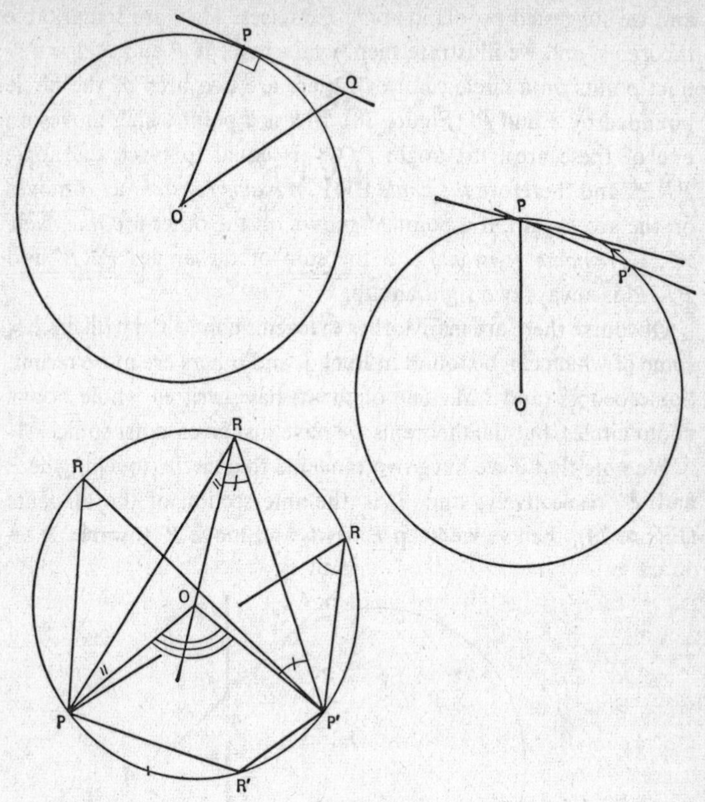

Figure 78

point P is at right angles to OP, where O is a fixed point. If we wish to introduce a concept which produces a tangent to other curves, besides the circle, we take a point P' on the circle near to P, and consider the chord, that is the line PP'. As P' moves on the circle towards P, the chord PP' changes its position, and *in the limit* it becomes the tangent to the circle at P. This is the enlarged concept of *tangent* which produces a tangent to most curves at a given point P on the curve.

There are one or two theorems on circles we have already used, and we suggested proofs in some Exercises. They are remarkable theorems, and we illustrate them again here. If P and P' are distinct points on a circle centre O, there are two arcs of the circle bounded by P and P' (Figure 78). If R is a point which moves on one of these arcs, the angle POP' is equal to twice the angle PRP', and therefore *the angle* PRP′ *remains constant* as R moves on the arc. Again, if a point R' moves on the other arc, *the angle* PR′P′ *remains constant*, and the sum of the angles PRP' and $PR'P'$ is always two right angles.

Of course there are many other theorems connected with circles, some of which can be found in Euclid, and others are more recent. Some people (and I am one of them) have written whole books about circles, but the theorems we have just given must suffice.

We note that if we have two tangents to a circle, touching at P and P' respectively, and T is the intersection of the tangents (Figure 79), then as we keep P fixed, and move P' towards P on

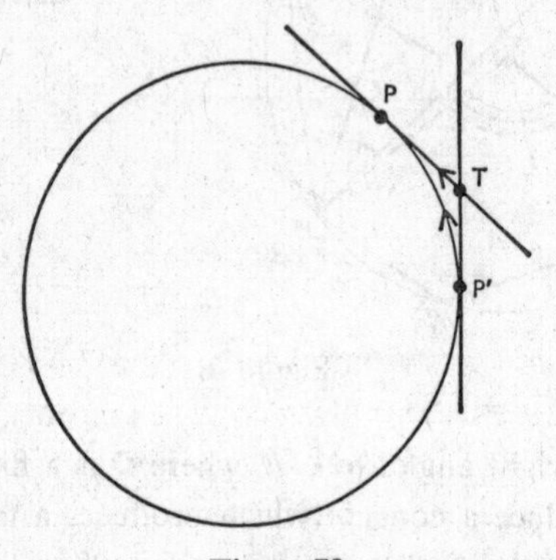

Figure 79

the circle, the point T of intersection of the two tangents *moves towards* P, the point of contact of the fixed tangent.

This is a property of tangents which is true for all the curves we shall be considering. One approach towards certain curves is to regard them as *envelopes* of their tangents. The simplest envelope is probably that traced out by a set of lines *at a fixed perpendicular distance from* a given point O (Figure 80). These lines *envelope* a circle, centre O, and if we draw enough of the lines, we can *see* the curve they touch, the *points* on the curve arising from the inter-

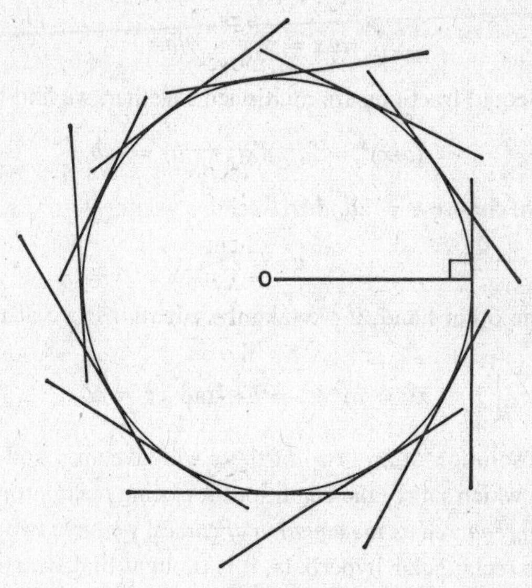

Figure 80

section of pairs of tangents which differ very little from each other. We call such tangents *successive* tangents, and we do not need to invoke the shades of Bishop Berkeley, who poured such scorn on Newton's use of infinitesimals, to acknowledge that we are not using mathematical language very precisely. But the reader will probably understand what we are doing, and that is the main point.

Later on in this chapter (p. 220) we shall be using sets of circles to obtain envelopes, but we now leave circles and investigate the *conic sections*. These first appeared in connection with the duplication of the cube (p. 30). The problem of solving the equation $a^3 = 2x^3$, which would enable a cubic altar twice as large as an original cube to be built, was replaced in Greek mathematics by the problem of *the insertion of two proportionals* between two given numbers a and b.

If x and y are the two proportionals, it was intended that

$$a/x = x/y = y/b.$$

If these equal fractions are multiplied together, we find that

$$(a/x)^3 = (a/x)(x/y)(y/b) = a/b,$$

and if we choose $a = 2b$, then

$$a^3 = 2x^3.$$

On the other hand, if we take the equations two at a time, we have

$$x^2 = ay, y^2 = bx, \text{ and } xy = ab,$$

and the solution of any two of these will give an x and a y, given a and b, which satisfy the conditions for being mean proportionals.

Since, *regarded as the equation of curves*, we have two parabolas and one rectangular hyperbola, it is thought that the study of the conic sections originated in these investigations.

The conic sections are aptly named because they can be obtained as sections of a right circular cone. A rubber ball and an electric torch shining on the ball will produce the conic sections as shadows. To the ancient Greeks the three types of conic section, ellipse, parabola and hyperbola, arose as sections perpendicular to a generator of three types of right circular cone. When the cone is acute-angled (Figure 81), the section gives an ellipse, when the cone is right-angled, a parabola, and when the cone is obtuse-angled, *one branch* of a hyperbola.

It took time to realize that one can take different sections of *the same* right circular cone, and obtain the three types of conic section, and that the hyperbola has *two* branches, which can be ob-

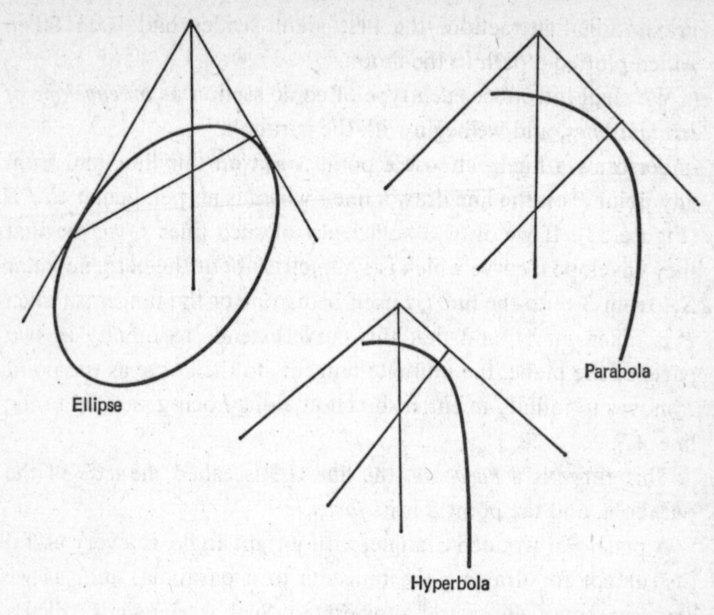

Figure 81

tained by considering cones as *extending through* their vertex, and not terminating there.

Apollonius of Perga (*c.* 247–222 B.C.) left a great treatise on the conic sections, and in this work (T. L. Heath, Cambridge University Press, 1896) he shows that the circular cones used need not even be *right* circular. In other words, join any point V to the points of a circle, the point V not lying in the plane of the circle, and the resulting cone will provide all types of conic section, if the planes are chosen correctly.

Conic sections became an intrinsic part of our culture when Kepler discovered that the planet Mars travels round the sun in an ellipse, the sun being at one focus. When Sir Isaac Newton (as he subsequently became) showed, in *Principia Mathematica*, using pure geometry only, that Kepler's discovery could be *deduced* as a mathematical consequence of the inverse square law of gravitational attraction, the first giant strides had been taken which plotted a path to the moon.

We shall introduce each type of conic section as *an envelope of straight lines*, and we begin with the parabola.

We draw a line l, choose a point S not on this line, and from any point P on the line draw a line t which is perpendicular to PS (Figure 82). If we draw a sufficiency of such lines t, we see that they envelope a curve which is symmetric about the perpendicular SA from S onto the line l, l itself being one of the tangents (when P is taken at A), and that this curve extends to infinity in two parts of the plane, the ultimate tangents to the curve as the point P moves to infinity in either direction along l being parallel to the line AS.

This curve is a *parabola*, the line AS is called the *axis* of the parabola, and the point S is its *focus*.

A plastic or wooden triangle, with a right angle, is a very useful instrument for drawing the tangents to a parabola, and, as we shall see, to an ellipse and a hyperbola. Such a triangle is called a *set-square* in England, but this term is not known in the United States. However, such triangles exist. On the other hand, although it seems possible to buy stencils for ellipses in University bookshops in the United States, it does not seem possible to buy any kind of stencil for parabolas or hyperbolas.

Another method for constructing conic sections by means of an envelope of lines is to produce the lines by *folding paper* in neat creases. Draw a line m on a sheet of paper, mark a point S on the paper which is not on the line, and fold the paper so that m passes through S. When it does so, crease the paper carefully. If you do this a number of times, the creases will be as indicated in Figure 82. There is a simple connection between the two processes we have described, using a set-square, and paper-folding.

From Figure 82 we can obtain a fundamental property of the parabola, considered as a set of points, from the envelope construction. Let R be the intersection of the tangents drawn from

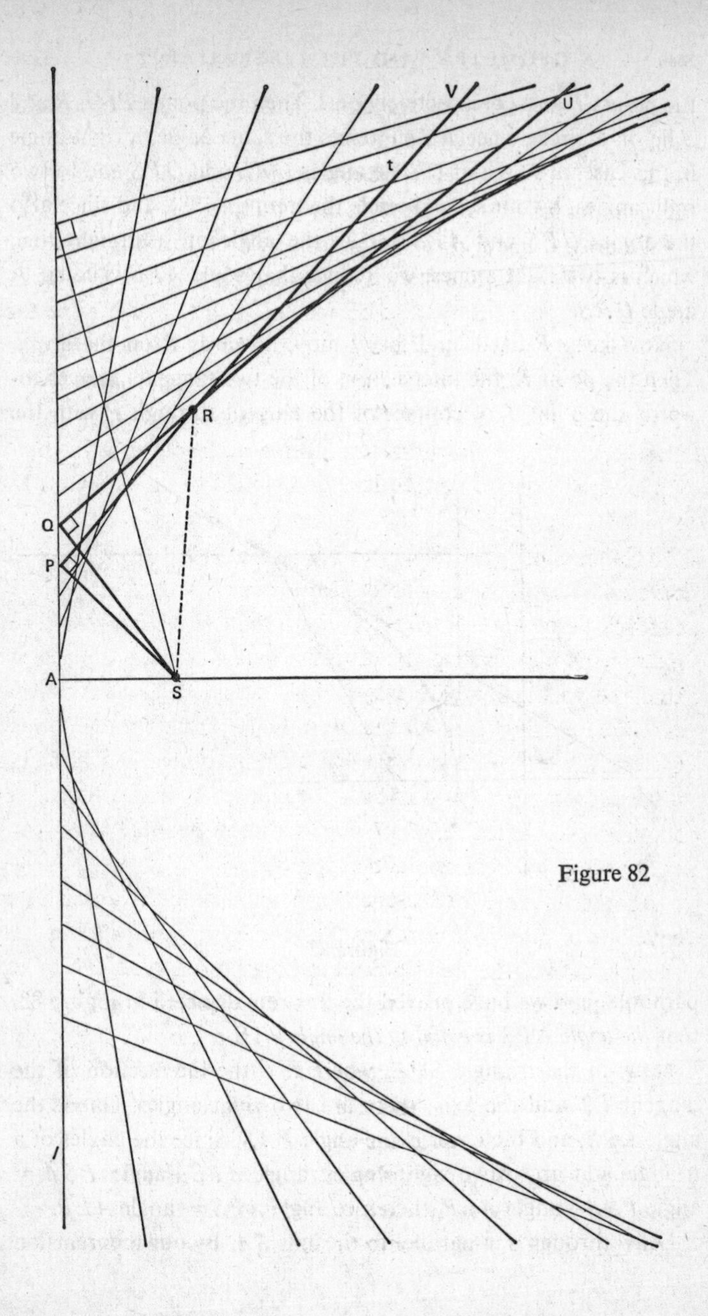

Figure 82

the points P and Q respectively on l. Then the points P, Q, R and S lie on a circle, since RS subtends the same angle, a right angle in this case, at P and at Q. The angles QRS and QPS add to two right angles, by our second circle theorem (p. 198), and since also the angles QPS and APS add to the angle on a straight line, which is two right angles, we deduce that angle APS is equal to angle QRS.

Now keep P fixed, and let Q move towards P on the line l. Then the point R, the intersection of the two tangents, moves towards the point T of contact of the tangent through P with the

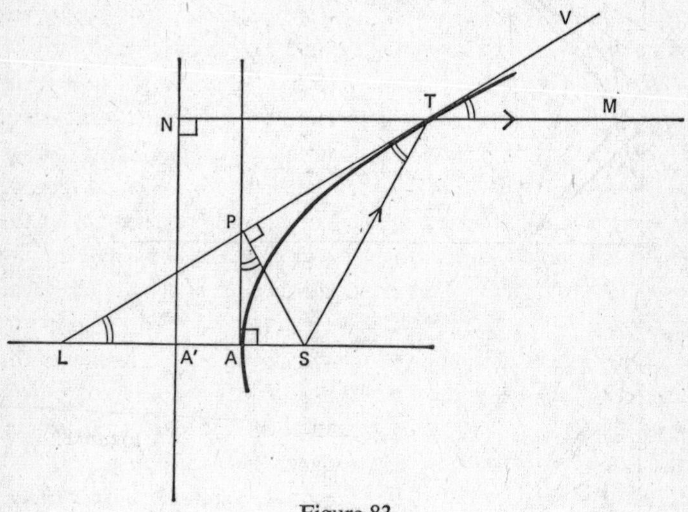

Figure 83

parabola, and we have proved the theorem depicted in Figure 83, that *the angle* APS *is equal to the angle* PTS.

Now, in the triangle SPL, where L is the intersection of the tangent PT and the axis, there are two right angles. One is the angle LPS, and the other is the angle PAS. Since the angles of a triangle add up to two right angles, angle APS+angle PSA = angle PSA+angle ALP, therefore angle APS = angle ALP.

Draw through T a parallel to the axis SA. By our theorems on parallel lines (p. 157), the angle VTM is equal to the angle ALP, and so angle ALP = angle APS = angle PTS = angle VTM, and from angle PTS = angle VTM we have one of the fundamental properties of the parabola, which shows that if our parabola is silvered, and we have a light source at S, then *any ray* leaving S is reflected at the mirror into a ray *parallel to the axis of the mirror*.

This is true, as we saw on p. 134, for only *a small part* of a spherical concave mirror, and parabolic mirrors are used for searchlights, car headlights, and, of course, radar, since a beam of rays or waves parallel to the axis of a parabolic mirror (or antenna) is reflected so as to pass through the focus S.*

Finally, if A is the midpoint of SA', the line through A' at right angles to the axis is an important one. It is called the *directrix* of the parabola, and if N is the foot of the perpendicular from T onto the directrix, it is easily proved that $ST = TN$, so that we now have the celebrated method of drawing a parabola as the locus of a point which moves so that *its distance from a fixed point* (the focus) *is equal to its distance from a fixed line* (the directrix).

In our paper-folding envelope construction, the line m is the directrix of the parabola. This focus-directrix definition of a parabola leads to a practical construction by means of nails, string, and so on, which we leave to the ingenuity of the reader.

The parabola is a curve, like the circle, which has *a fixed shape*, and its size is dependent on the value of one variable, which is usually taken as the distance SA.

If a stone is thrown at an angle, that is, not vertically upwards, and air resistance is negligible, the path traced out by the stone (the trajectory) is a parabola, with a vertical axis, and the distance SA is proportional to the horizontal component of the initial velocity of the stone. If we determine the height of the directrix of

*Dürer gives a sketch of a parabolic *Brenspiegel* (burning glass) in the *Underweysung*, showing parallel rays coming to a focus.

the path above the point of projection, it turns out to be equal to the height from which a stone would have to fall vertically to acquire the initial velocity of projection of the stone, which is not, however, projected vertically.

Now, suppose that we have a fountain with a number of jets, and that from each jet water issues with the same speed. Then we shall see a number of parabolas, and these parabolas also have an envelope. This envelope turns out to be a parabola also, with focus at the common point of all the parabolas, and the tangent at the vertex of the envelope is the common directrix of all the component parabolas. A fountain of this sort is a lovely spectacle, but modern fountains usually consist of one aimless jet gushing all over the place.

A proof of this envelope theorem is suggested in Figure 84, and follows quite simply from the theorems already proved.

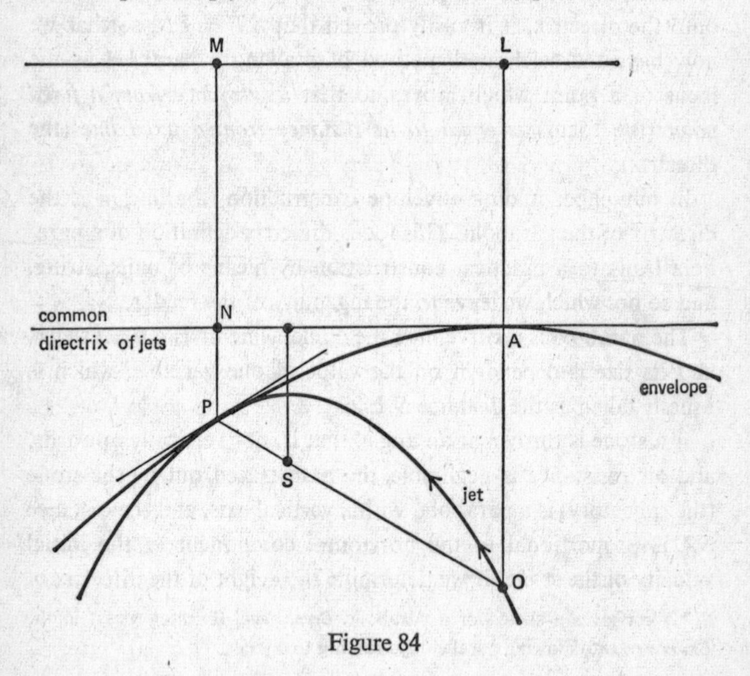

Figure 84

A less aesthetic application of the envelope theorem is to the science of ballistics. If a gun has a given muzzle velocity, its projectiles cannot reach outside the enveloping parabola we have just described, in any vertical plane through the gun. To the enemy this parabola is the parabola of safety, of course. It is quite simple to show that the maximum range for a given velocity of projection is obtained on an inclined plane through the point of projection O if the initial direction of projection bisects the angle between the vertical through O and the inclined plane, so that as a special case, if you want to throw a ball as far as possible on a flat field, throw it at an angle of 45° with the horizontal.

We now turn to the ellipse, and show how this curve can be obtained as an envelope. We draw a circle, centre C, and choose a point S inside it (Figure 85). Join S to any point Q on the circumference of the circle, and draw a line through Q perpendicular to SQ. Again, a right-angled set-square is helpful. Then the set of all such lines envelopes an *ellipse*.

We can obtain some properties of the ellipse fairly rapidly. If the perpendicular through Q to QS intersects the circle again at R, and we draw a perpendicular through R to RQ, this intersects SC at a point S' such that $SC = CS'$, so that S' is also a fixed point.

We now repeat the procedures we used with the parabola. Let Q' be a point on the circle near to Q, and consider the perpendicular through Q' to SQ'. This is another tangent to the ellipse. Let the tangent emanating from Q and Q' intersect at T. Then the points S, Q, Q' and T lie on a circle, and so the angle UQS is equal to the angle STQ'.

As Q' approaches to Q on the circle, the line QQ' becomes the tangent to the circle at Q, and the point T approaches to the point P *of contact* of the tangent QR with the ellipse. So we end up with the angles shown equal (Figure 86), and since the argument can be repeated, using S' instead of S, and the angles WQP and WRP are equal for a circle, we end up with the theorem that if S

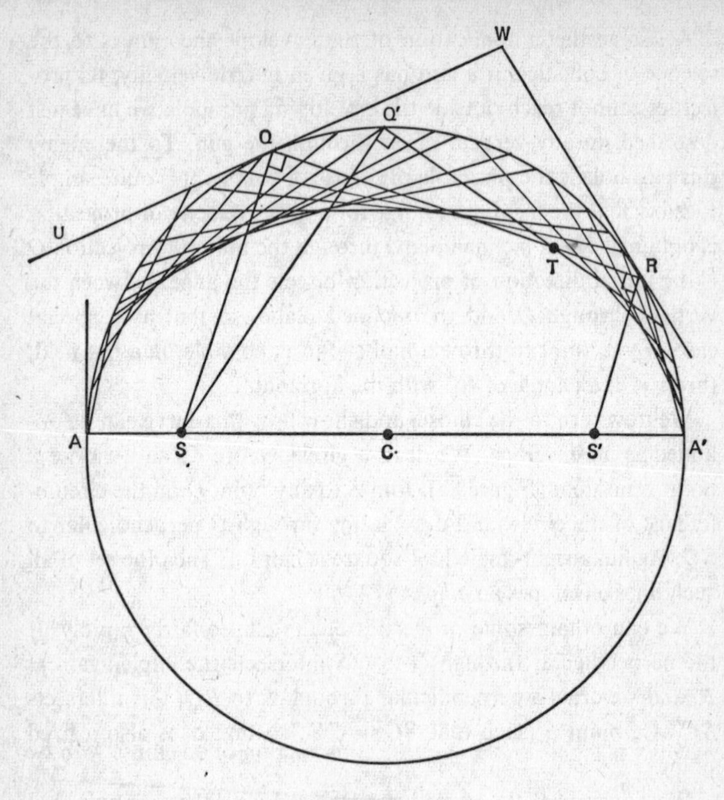

Figure 85

and S' are the two foci of an ellipse, and P is any point on the ellipse, then the angles between the tangent to the ellipse at P and the lines SP and $S'P$ are equal.

If we have an elliptical mirror, a ray of light emanating from S will, after reflexion, pass through S'. There exist elliptical billiard tables which exhibit this property for billiard balls, although the laws of reflexion in such a case are not exactly those of light.

If we keep S fixed, and move the centre C of the circle off to infinity, the ellipse will look more and more like a parabola, focus

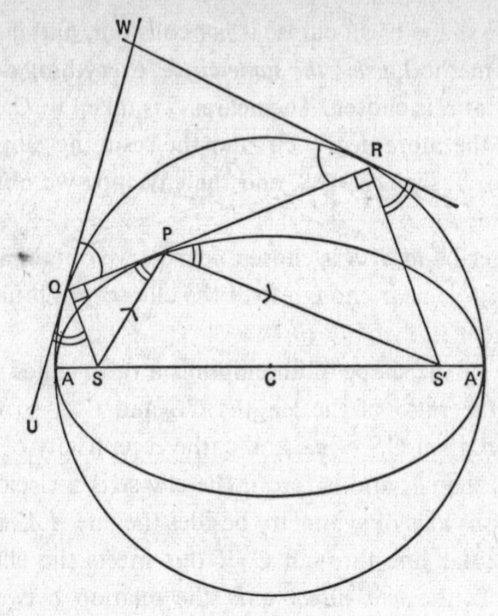

Figure 86

S, and the ray of light reflected at *P* will end up parallel to the axis.

We can prove quite simply that $SP+S'P = AA'$, and this gives a practical method for the construction of an ellipse with two nails and a length of string (Figure 87), the nails being the foci,

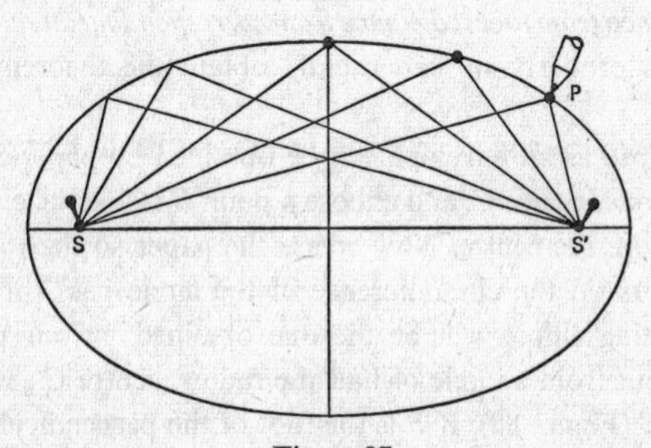

Figure 87

S and S'. The shape of an ellipse is *not* constant, and if we pursue the envelope method, *using the same circle*, everything depends on where the point S is chosen. The nearer S is taken to C, the centre of the circle, the more nearly circular the resulting ellipse. If S is taken at C itself, then $S' = S$, and the envelope we obtain is the given circle.

On the other hand, if S is chosen near A, we obtain a very thin ellipse, and if S is near enough to A the ellipse will look very like the line segment AA'.

A measure of the shape of an ellipse is a ratio called the *eccentricity*. It is the ratio of the lengths CS and CA, so that if we write $CA = a$, then $CS = ae$, and e, the eccentricity of an ellipse is always less than 1, and is zero if the ellipse is a circle. There is another obvious axis of symmetry besides the line AA', and this is the perpendicular line through C. If this meets the ellipse in B, and if $CB = b$, the *semi-minor axis*, the relation between a and b is $b^2 = a^2(1-e^2)$.

The planets, which move in ellipses with the sun at one focus, have a small eccentricity.

There is also a focus-directrix method for constructing an ellipse. If we draw a perpendicular to the major axis AA' at a point N where $CN = a/e$, this line is the directrix which corresponds to the focus on the same side of C. For any point on the ellipse, the ratio *distance from focus*:*distance from corresponding directrix* $= e$. From this property we can rapidly obtain the theorem: $SP+S'P = 2a$.

The ellipse as an envelope can be obtained by paper-creasing. Draw a circle, centre C, and choose a point S inside the circle, but distinct from the centre. Now crease the paper so that, one at a time, points on the circumference of the circle pass through S. The resulting ellipse will be the one obtained by our previous method, but from a circle of half the radius, centre C', where C' bisects SC (Figure 88). If P is the foot of the perpendicular from S onto the crease, it is not difficult to show that *the locus of* P *is a*

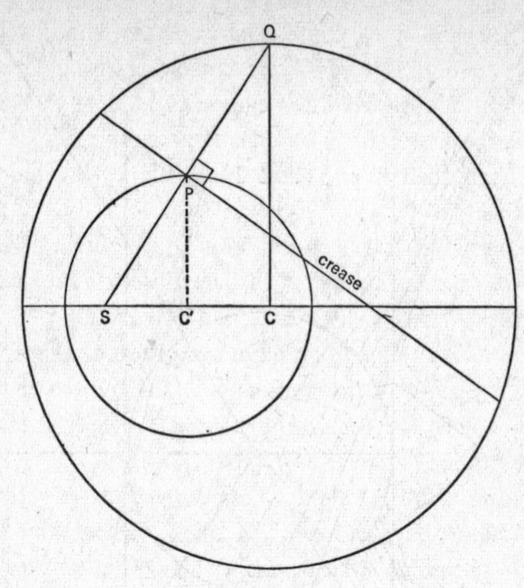

Figure 88

circle, centre C' and radius half the radius of the original circle. Since, if we reduce the original circle and all the points and lines appertaining to it by one half, we obtain the circle centre C', and the same point S, since $SC = 2SC'$, the ellipse obtained by paper-folding is the same shape as the original ellipse, but half the size.

Our final conic section is the hyperbola, obtained, as one might now anticipate, by drawing a circle centre C, choosing a point S *outside* it, joining S to points Q on the circumference, and drawing lines through Q at right angles to SQ (Figure 89).

The resulting envelope is markedly different from the two previous ones we have obtained. It separates into *two curves*, or *branches* of the one hyperbola.

From S we can draw two tangents SM and SN to the circle, and we see that the points between M and N on the arc of the circle facing S give tangents to one branch of the hyperbola, and

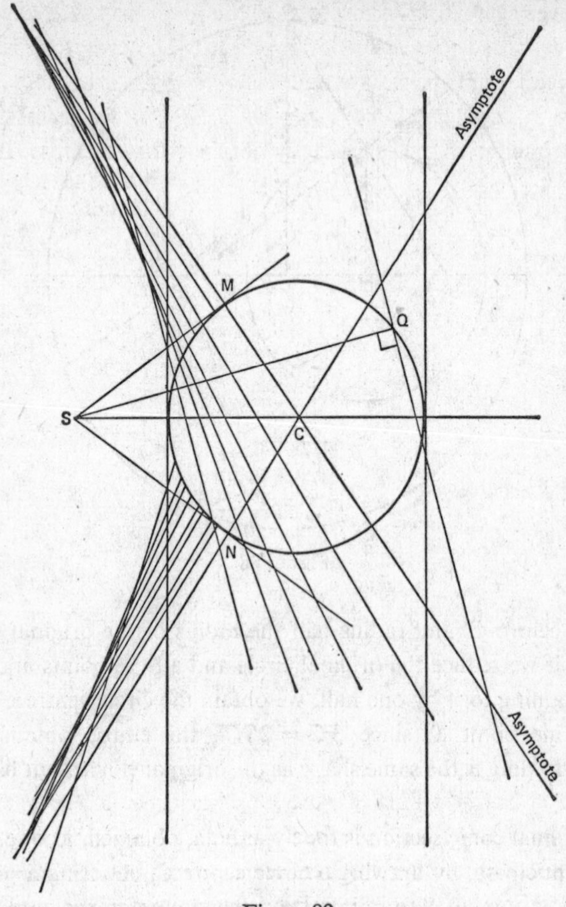

Figure 89

the points on the other arc of the circle give tangents to the other branch.

The tangents to the hyperbola at M and N both pass through the centre C of the circle, since MS is perpendicular to CM, and NS to CN. These special tangents to the hyperbola are called *the asymptotes*. As a point on the hyperbola moves off to infinity, it also moves nearer and nearer to an asymptote, and the asymptotes can be regarded as *tangents at infinity*. If the asymptotes are at right angles, the hyperbola is called a *rectangular* hyperbola.

There is another focus S', besides S, and we easily obtain a theorem similar to the one already obtained for an ellipse. If P is

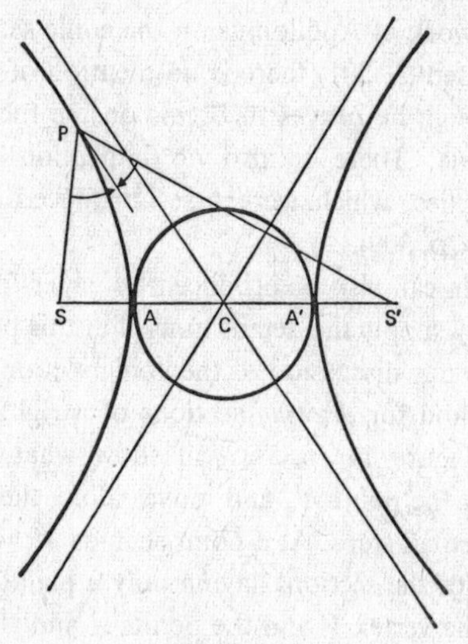

Figure 90

a point on the hyperbola (Figure 90), *the tangent at* P *to the curve bisects the angle between the focal distances* SP *and* S′P.

There is also another theorem corresponding to the *sum* of the focal distances of an ellipse being constant, but in the case of an hyperbola the *difference* is constant. If P is as shown, then $S'P-SP$ is constant while P remains on its branch of the hyperbola. If P is on the other branch, then it is $SP-S'P$ which is constant, the constant in each case being AA'.

For the hyperbola $CS = eCA$, where $e > 1$, and each focus has a corresponding directrix at a distance CA/e from the centre C. The hyperbola can also be traced as the locus of a point which moves so that its distance from a fixed point is always equal to e times its distance from a fixed line, where $e > 1$.

We see that the focus-directrix definition of the conic sections is a unifying one, the change in the value of the eccentricity e through the value $e = 1$ produces the changes from ellipse, through the parabola to the hyperbola.

In the great work of Apollonius on the conic sections we have already mentioned (p. 201) there is no mention of the focus of a parabola, although he proves theorems on the foci of an ellipse and a hyperbola. There is also no indication of the focus-directrix properties, which were first introduced by Pappus of Alexandria (*c.* A.D. 300).

The hyperbola can also be produced by paper-folding, but the reader can easily supply the details himself at this point.

We complete our discussion of the conic sections by returning to Dürer's method for *drawing* sections of a right circular cone (p. 58). Here Figure 14 (p. 59) will show what is happening. Dürer begins at the point A', and moves along the major axis of the sections in equal steps. At a point such as M, we wish to find the width MQ of the section, having only a plane section of the cone through the vertex V and the points A and A', as shown in Figure 91.

If we take a plane through PQ perpendicular to the axis of the cone, this plane will cut the cone in a circle through P and Q, and we can obtain the diameter of this circle by drawing a line through the point M in the plane section VAA' perpendicular to the axis of the cone, and finding where this line intersects the two generators VA and VA' of the cone in the section (Figure 91). In *this* circular section PMQ *is a chord through* M *perpendicular to the horizontal diameter of the circle*, and MQ is easily found by drawing, once we have the circle.

Dürer aligned the major axis of the section vertically alongside the section of the cone, and marked off the various widths MQ as he found them, but this is not shown in his drawing (Figure 14),

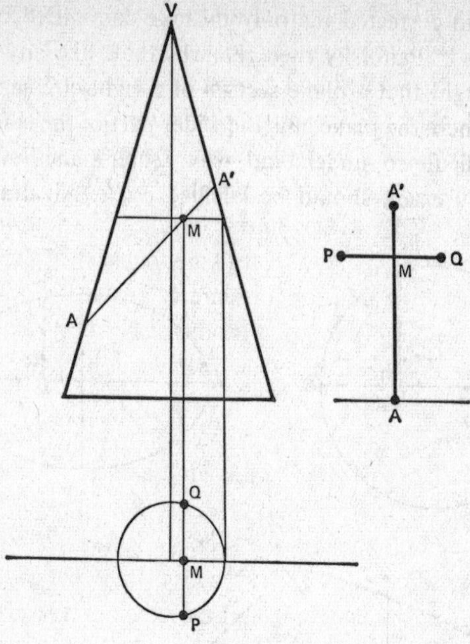

Figure 91

and some commentators have completely misunderstood what he was doing. In fact Dürer was following the method by which Apollonius had shown that plane sections of a circular cone are ellipses, parabolas or hyperbolas.

Attempting the actual drawing by Dürer's method is not too easy, and even a small slip of the ruler or compasses will produce an error of the kind which amused Kepler, an egg-shaped ellipse. On the other hand Dürer did use the term *Eierlinie* for the curve instead of the Latin *Elipsis*, on the grounds '*dasz sie schier einem Ei gleich ist*'.

In his great book on Dürer, Erwin Panofsky says: 'Crude though this is, this method, which may be called a genetic as opposed to a descriptive one, announces, in a way, the procedure of analytical geometry, and did not fail to attract the attention of

Kepler, who corrected the only mistake committed by Dürer in his analysis.'* Panofsky then remarks that, like any schoolboy, Dürer thought that a plane section of a right circular cone must be wider where the plane cuts the wider part of the cone.

It is difficult to understand why Dürer's method, which is theoretically exact, should be labelled *crude*. All drawing is, of

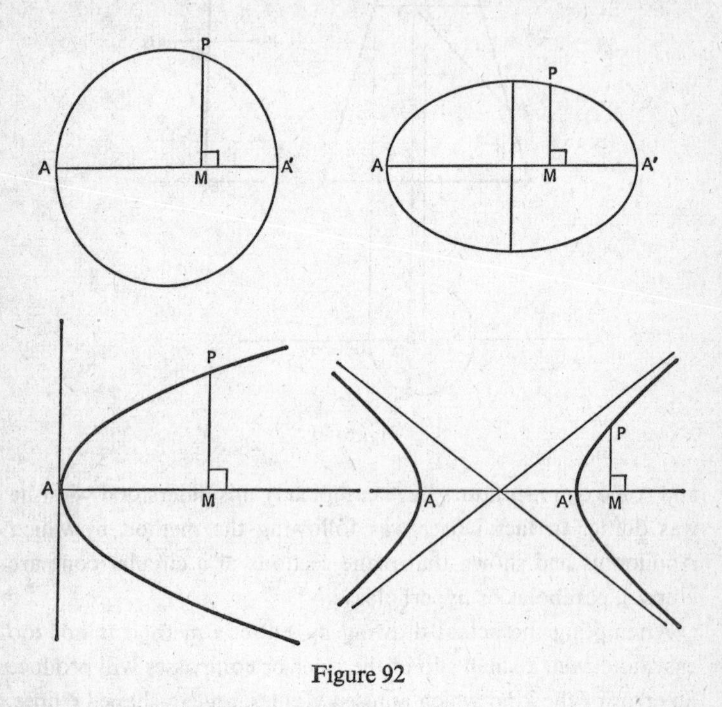

Figure 92

course, crude from the theoretical point of view. As we have remarked, Dürer was faithfully reproducing Apollonius, who had shown that the three sections of a circular cone may be defined as follows (Figure 92):

For an ellipse, A and A' are fixed points, P a variable point on

* *The Life and Art of Albrecht Dürer*, Princeton University Press, 1943, p. 225.

the curve, PM is perpendicular to AA', then, k being a positive constant,

$$(PM)^2 = kAM.MA'.$$

If the ellipse is a circle, $k = 1$; for a hyperbola, k is a negative constant. For a parabola, A is a fixed point, AX a fixed line, and PM is perpendicular to AX. Then $(PM)^2 = kAM$, where k is a constant. Analytic geometry was implicit in this work of Apollonius.

Dürer's methods have never appeared in any ordinary English book on geometry, as far as I know, but I am told that they can be found in German textbooks. We apply Dürer's method to obtain a drawing of a parabolic section of a right circular cone, and we shall also obtain its *equation*, the connection between the length $A'M$ and the length MQ.

To simplify the calculations, take the cone as one with semi-vertical angle 30° (Figure 93). The plane section has to be parallel to a generator, if we wish to obtain a parabola. Suppose that A' is at a distance a from the vertex V of the cone. To obtain the width of the parabolic section at a point M distant x from A', we draw through M a perpendicular to the axis in the section shown, meeting the two generators in R and S. Draw the circle with RS as diameter, and mark the point M on the diameter, drawing a chord QMP perpendicular to this diameter RS.

Then $MQ = y$ is the width of the parabola at the point M on its axis. Since the angle $SVR = 60°$, the triangle MRA' is equilateral, so that $MR = MA' = x$, and we also have $SM = VA' = a$, triangle SRV being equilateral. By the theorem of circle geometry just quoted, $MS.MR = (MQ)^2$, and so we have the relation $ax = y^2$, and this is the well-known equation for a parabola.

A drawing of a hyperbolic section, using Dürer's method, is also given in Figure 94, but only the points on one branch have been calculated.

It may seem curious to switch from the conic sections to epi-

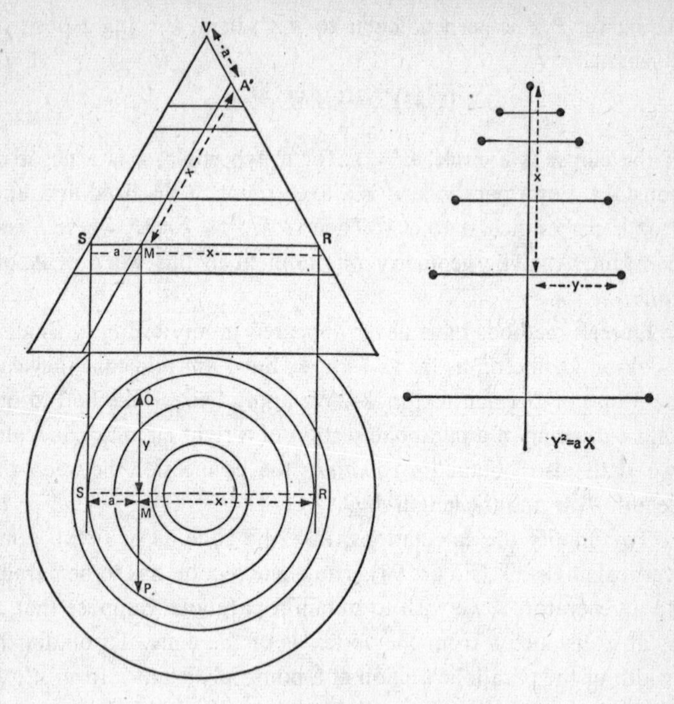

Figure 93

cycles, but the conic sections usurped the leading role played by epicycles in astronomy from the moment Kepler, as the result of fantastically laborious calculations, determined the path of the planet Mars as an ellipse, with a small eccentricity, with the sun in one focus. The circle had been regarded for millennia as the perfect curve, so it was natural to feel certain that the planets travelled in circular paths. When this did not fit the phenomena, *epicycles* were invented, circles moving on circles.

To be more precise, suppose that the point Q describes a circle, centre O and radius a (Figure 95), and the point P, at the same time, describes a circle, centre Q and radius b. We assume that Q moves in a uniform way around its circle, and that to an observer

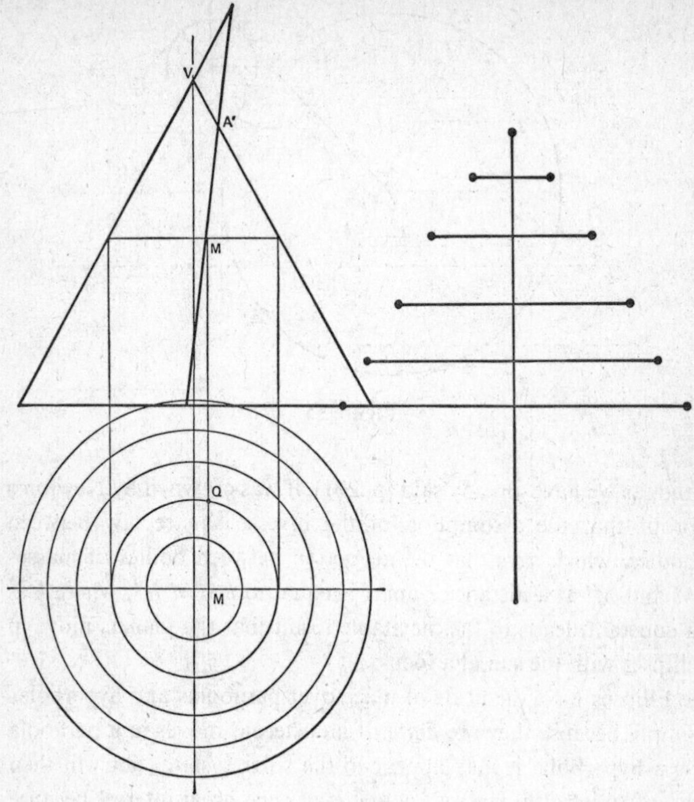

Figure 94

carried along with the point P it seems as if the circle centre Q is also being described uniformly. Then the path described by P in the plane is called an *epicycle*, and the motion can become more involved if we imagine a further point R describing a circle with P as centre and of radius c, also uniformly.

More and more complicated epicycles were invented *to save the phenomena*, that is, to produce a curve which fitted in with planetary motion, when previous epicycles were found to be defective. Kepler's discovery was a great simplification, of course,

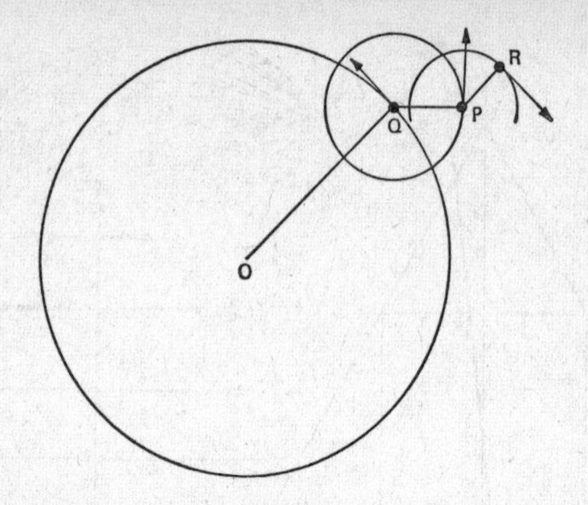

Figure 95

and, as we have already said (p. 201), it was crowned by Newton's proof that the assumption of the inverse square law between bodies, which says that the attraction between bodies of masses M and M' at a distance r apart is equal to kMM'/r^2, where k is a constant, leads to the inevitable result that the planets move in ellipses with the sun at a focus.

Ellipses have the pride of place over parabolas and hyperbolas simply because they are *finite*. If an asteroid moves in a parabola or a hyperbola, it may appear in the solar system, but will then go off to infinity. Halley's comet is of such great interest because Edmund Halley calculated its path as an elongated ellipse, and forecast its reappearance. It was last seen in 1910 and is expected to reappear in 1986.

Epicycles will be discussed again later, but first we consider some beautiful curves which can be obtained as *an envelope* of circles. We begin with the cardioid (Figure 96).

Draw a circle (which we may call the base-circle), and mark a point A on it, which we keep fixed. With centre at any point Q of the circle, and radius QA, draw another circle, and repeat this for a large number of positions of Q, spread uniformly around the

Figure 96

base-circle. The heart-shaped curve which all these circles envelope is the cardioid. The point A, where two portions of the cardioid touch, is called a *cusp*.

Part of the fascination of the curves we shall discuss, already exemplified with the conic sections, is the possibility of their *multiple generation*. To obtain the cardioid as a locus of *points*, we draw a circle once more, of diameter a, fix a point A on it, join the fixed point A to a variable point Q of the circle, and produce AQ by the fixed length a to P (Figure 97). Then the locus of P is a cardioid with cusp at A.

Note that as the line AQ rotates about A, the point Q will eventually be found on the lower half of the circumference of the

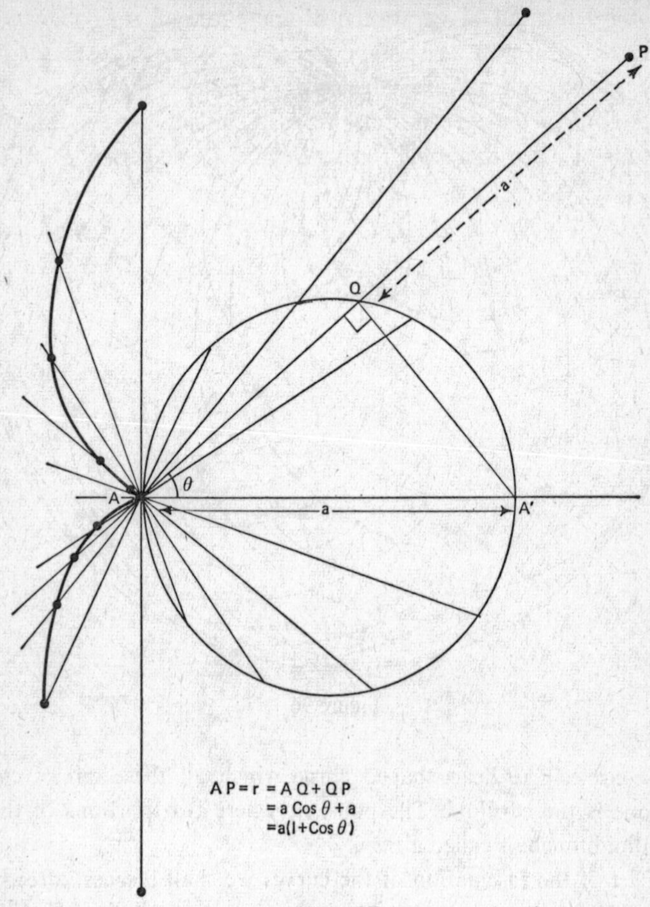

Figure 97

fixed circle, and the length AQ must be *subtracted* from the fixed length a.

For the record, if the angle PAA' is called θ, where AA' is the diameter of the circle, and the length AP is called r, then the *polar* equation (that is the (r, θ)-equation) of the cardioid is

$$r = a + a\cos\theta.$$

A third method of generation of the cardioid is as follows. Again we begin with a circle and a point A on it, and consider a *tangent* to the circle at a variable point Q on the circle. Now let P be the foot of the perpendicular from A onto the tangent at Q (Figure 98). Then the point P describes a cardioid with cusp at A

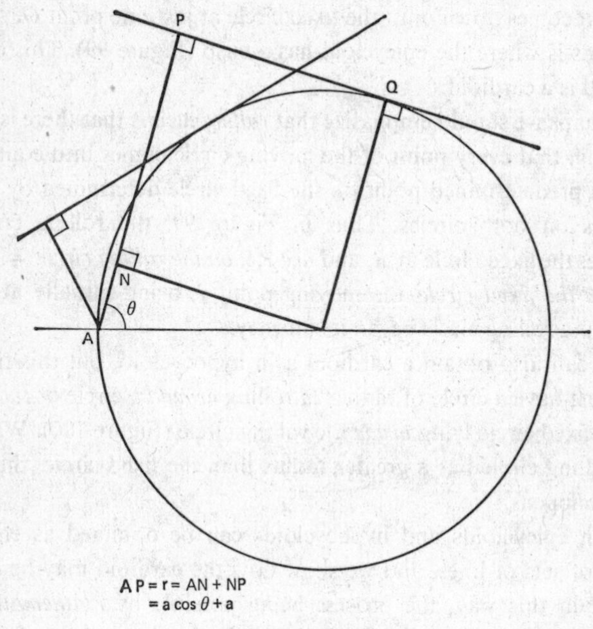

Figure 98

as Q varies. We say that the point P describes *the pedal curve* of the point A with respect to the circle, so the cardioid appears as a pedal curve.

We shall be discussing circles rolling on fixed circles, and we give some definitions. If a circle C rolls on the *outside* of a fixed circle C', the curve described by a fixed point P on the circumference of the rolling circle C is called an *epicycloid*. This curve is evidently a special form of *epicycle*.

If the curve described by P is obtained by a circle C rolling on the *inside* of a fixed circle C', it is called a *hypocycloid.*

It is easy to see that when the rolling circle brings the point P on the circumference of C towards the circumference of C', and then up and away again, the moving point P *traces out a cusp.* If a circle of radius a rolls outside an equal circle of radius a, the point P comes down onto the fixed circle at just one point O, say, and this is where the epicycloid has a cusp (Figure 99). This epicycloid is a cardioid.

Perhaps we should emphasize that *rolling* means that there is no slipping, that every point of the moving circle comes into contact with a predetermined point on the fixed circle determined by arc lengths on both circles. Thus in Figure 99, the rolling circle touches the fixed circle at R, and *arc* RP *on the rolling circle = arc* RO *on the fixed circle*, the moving point P being initially at O, when the rolling circle began its journey.

We can also obtain a cardioid as a hypocycloid, but this time we must have a circle of radius $2a$ rolling *around* a circle of radius a, the fixed circle lying *inside* the rolling circle (Figure 100). When the rolling circle has a greater radius than the fixed circle, this is what happens.

Both epicycloids and hypocycloids can be obtained as envelopes of sets of lines, and we show how the cardioid may be obtained in this way, the process being possible by *mathematical embroidery*, various coloured threads being used instead of drawn lines.

Divide a circle (Figure 101) into equal arcs subtending 10° at the centre. This will give 36 divisions. Mark one of the divisions as the point A, and the opposite end of the diameter as the point A'. Beginning at A, go round the outside of the circle and mark the divisions as 0, 1, 2, . . ., the point A being marked 0, so that A' is marked as 18, and eventually A will be reached again and marked as 36. Now start at A', and on the *inside* of the circle, the sense of rotation being as before, mark intervals of 20°, the

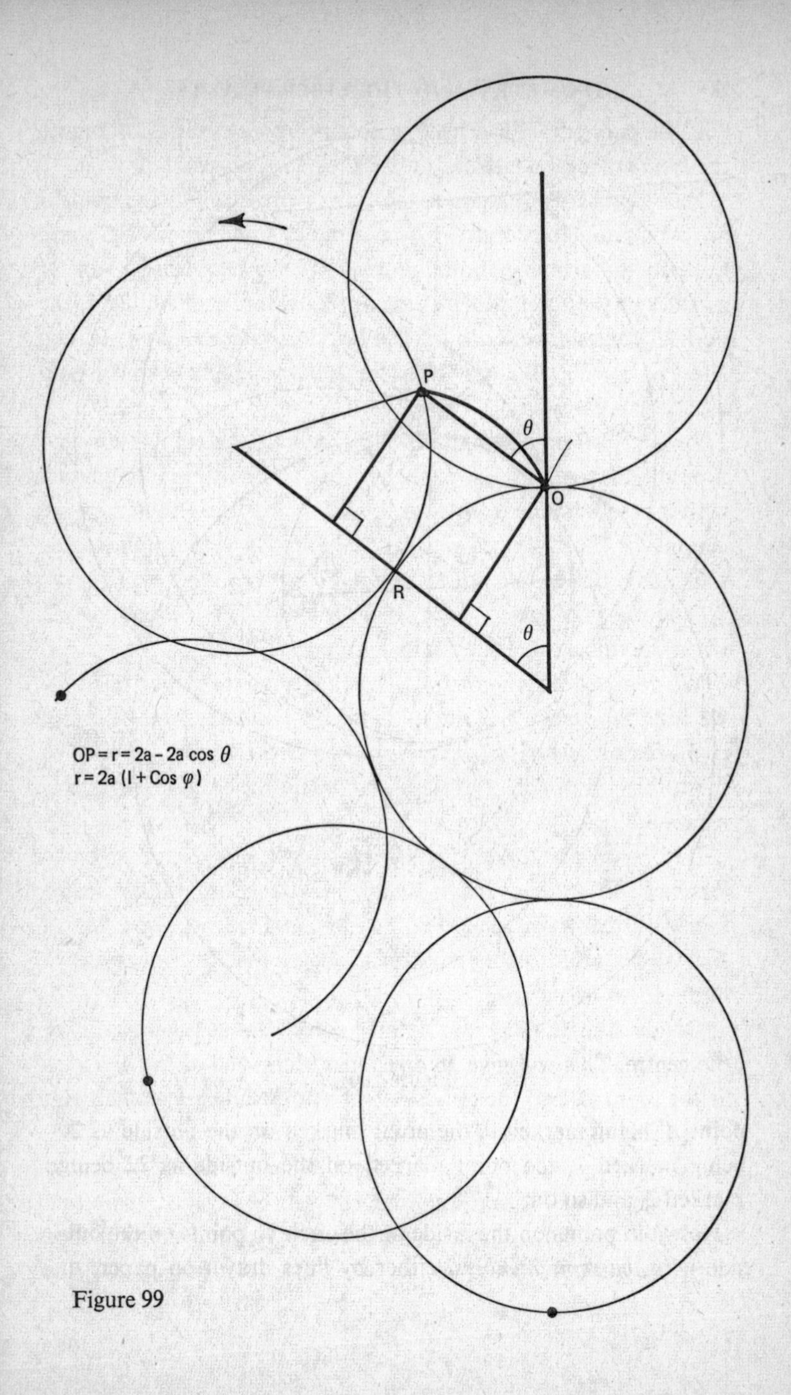

Figure 99

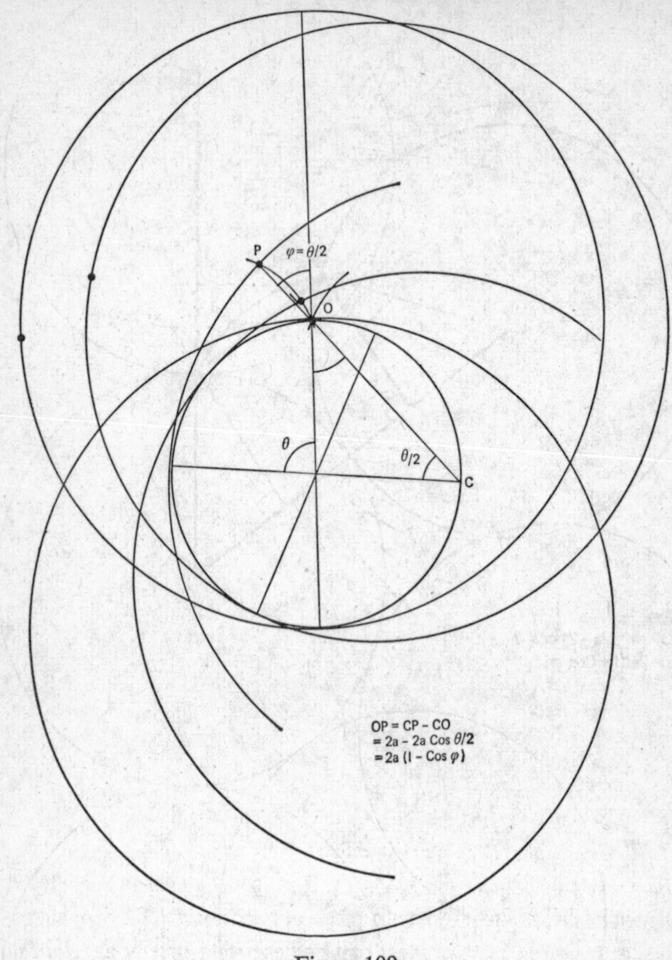

Figure 100

point A' being marked 0, the point marked on the outside as 20 being marked 1, the point marked on the outside as 22 being marked 2, and so on.

Now join points on the inside of the circle to points on the outside *with the same numbers*, either by lines drawn on paper, or

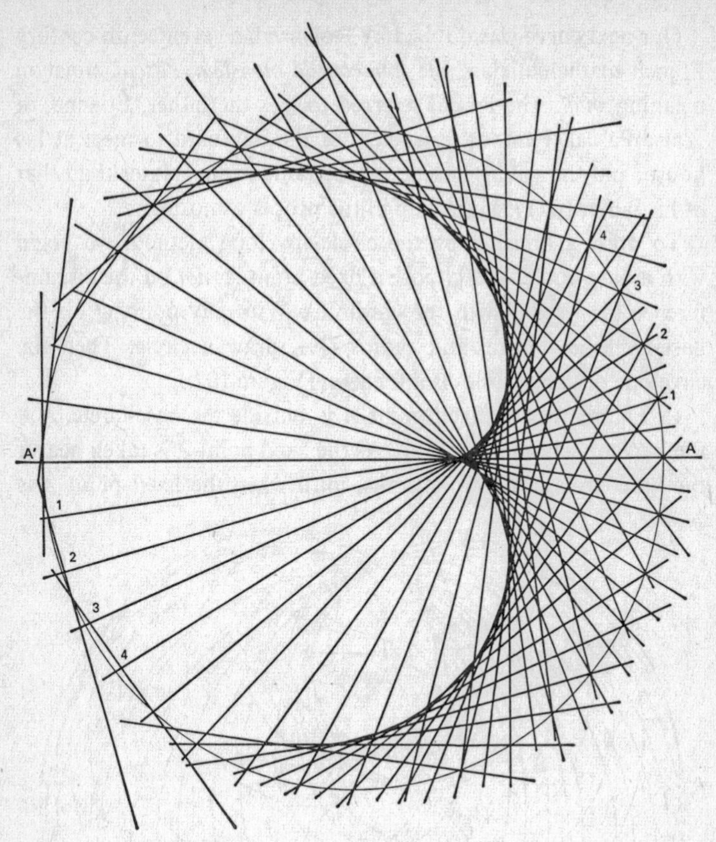

Figure 101

with stitches in fine thread on a piece of cardboard. The envelope is the cardioid, and the process can be described briefly as: *Join* n *to* 2n, *same orientation*. To obtain a complete figure, the numbering will have to be continued past the original starting-points, 0 being also marked 36, 1 being marked 37, and so on.

The name *cardioid* (heart-shaped) was first used by de Castillon in *The Philosophical Transactions of the Royal Society* of 1741, but he was not the first to consider this attractive curve.

Our next curve was dubbed by Roberval, a seventeenth-century French mathematician, the *limaçon de monsieur Pascal*, limaçon meaning *snail*. The Pascal referred to was the father, Étienne, of Blaise Pascal. Famous geometers of the day used to meet at his house, and the young Pascal, whose fame is far greater than that of his father (p. 179), grew up in the proper atmosphere.

To draw a limaçon by the circle-envelope method, we begin with a base-circle, and choose a fixed point A, *not* on the circumference (as we did with the cardioid). With any point Q on the base-circle as centre, and radius QA, draw a circle. Then the envelope of these circles is a limaçon (Figure 102).

In Figure 102 the fixed point A is outside the base-circle, and the limaçon has an inner loop. As the fixed point A is taken nearer the base-circle, the loop shrinks, until when the fixed point A is

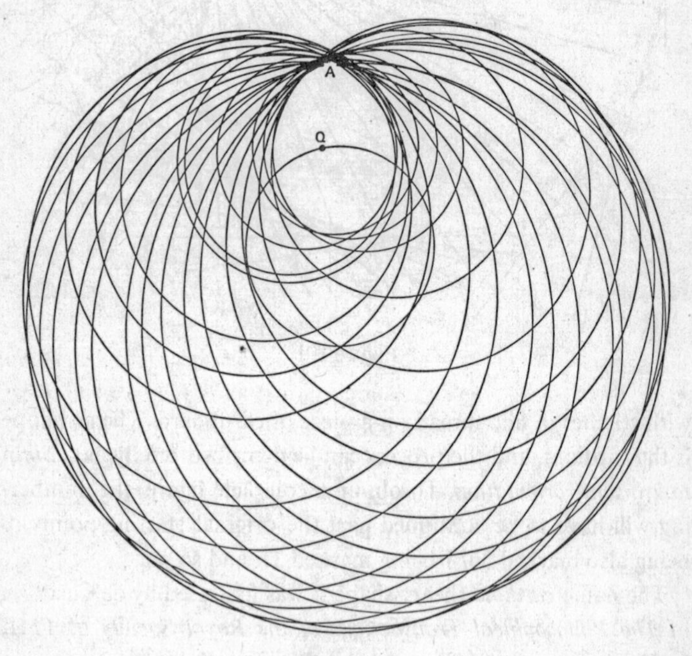

Figure 102

Figure 103

taken *on* the base-circle, we obtain the cardioid. When the fixed point A is taken *inside* the base-circle (Figure 103), we obtain a limaçon without an inner loop.

The limaçon can be obtained as a point-locus in at least two ways. Once again, we draw a circle of diameter a, fix a point A on it, and join A to a variable point Q on the circle, producing AQ to P by the fixed length $k \neq a$. Then the locus of P is a limaçon.

If $k = a$ we obtain the cardioid. Once again, for the record, the polar equation of the limaçon is

$$r = k + a\cos\theta.$$

To obtain the limaçon as the pedal of a circle, we take the fixed point A *not* on the circle, and drop perpendiculars from A onto tangents to the circle. The locus of the feet of the perpendiculars is a limaçon.

As we now show, Dürer should have the credit for discovering this curve. In the *Underweysung*, published in 1525, to which we have referred in our chapter on Dürer a fair number of times, Dürer gives the following construction:

A circle is drawn and the circumference is divided into twelve equal parts, these being numbered like the hours on a clock-face (Figure 104). From the ends of the radii 1, 2, 3, . . . lines are

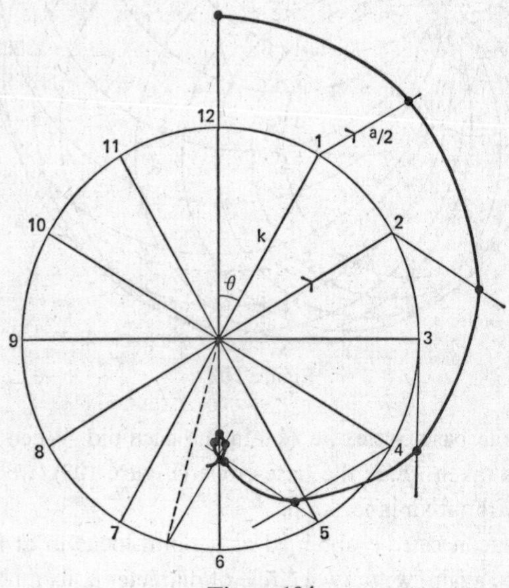

Figure 104

drawn, all of the same length, respectively parallel to the radii 2, 4, 6, . . . The ends of these lines are then connected. Finer divisions of the clock face, of course, produce more points on the curve.

Our assertion that Dürer's curve is a limaçon cannot be based on the mere *appearance* of the curve, and we either have to reconcile Dürer's method with one of the known constructions, or actually find the equation for his curve, and show that this is the same equation as that of a limaçon.

We find the equation, but first remark that Dürer was naturally interested in epicycles (p. 218), and in the *Underweysung* there is a picture of what may be called an *epicyclic ruler* (Figure 105). The above curve can actually be described by a point attached to a circle which rolls on the outside of another circle, the point *not* being on the circumference of the rolling circle. Such a curve is called an *epitrochoid*, and Dürer's construction enables us to establish that his curve is the particular epitrochoid which is a limaçon. But we shall actually find the equation, by obtaining the X and Y co-ordinates of a variable point, first for a known limaçon, then for Dürer's curve, showing that they are the same.

The polar equation of a limaçon is $r = k+a\cos\theta$, and if we move the origin of co-ordinates to the centre of the base circle, instead of

$$\left.\begin{aligned} x &= (k+a\cos\theta)\cos\theta, \\ y &= (k+a\cos\theta)\sin\theta, \end{aligned}\right\}$$

the equations become

$$\left.\begin{aligned} X+a/2 &= (k+a\cos\theta)\cos\theta, \\ Y &= (k+a\cos\theta)\sin\theta. \end{aligned}\right\}$$

If we take the $a/2$ over to the right-hand side, and use the known identity:

$$\cos 2\theta = 2\cos^2\theta-1,$$

the equations now become

$$X = k\cos\theta+\frac{a}{2}\cos 2\theta,$$

$$Y = k\sin\theta+\frac{a}{2}\sin 2\theta.$$

In Dürer's construction, if we take the *radius of the clock* to be k, and the fixed extension as $a/2$, measuring angles clockwise

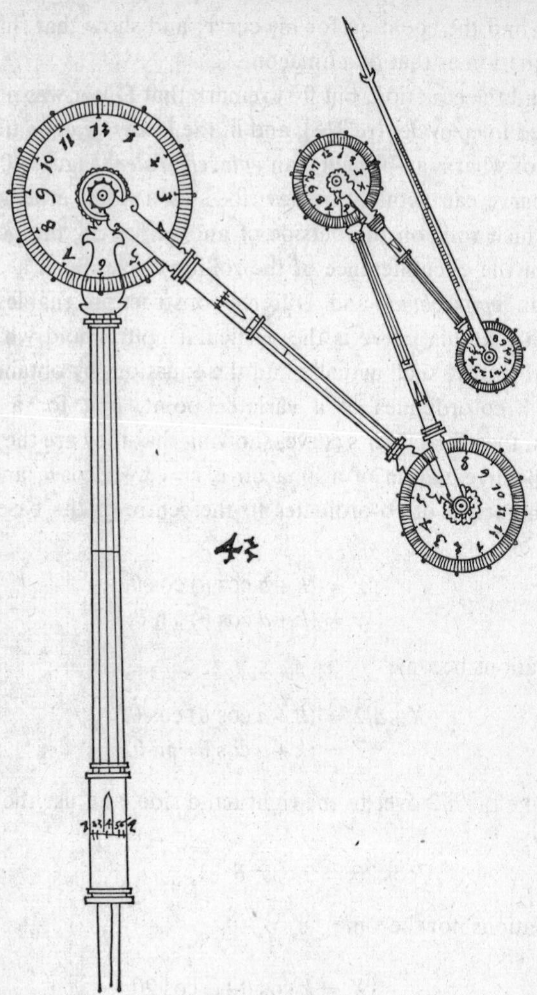

Figure 105

from the radius through 12 o'clock, we obtain precisely these equations for the X- and Y-co-ordinates of the end-point. The limaçon should certainly be called *Dürer's limaçon*.

We now consider the last curve for which we give a circle-envelope construction, the *nephroid*. This means *kidney-shaped*, and was the name given to this two-cusped epicycloid by Proctor, in 1878. Actually it is like a kidney sliced through the middle and opened up, preparatory to frying. If, like Leopold Bloom, in James Joyce's *Ulysses* (and myself), you like fried kidneys, you will also find the nephroid a pleasing curve. If you do not, the association may worry you, and the next curves await your consideration.

Draw a base-circle once more, and fix one of its diameters, the line l. Choosing any point on the base-circle as centre, draw a circle which *touches* the line l. The envelope of this set of circles is the nephroid (Figure 106).

The nephroid has two cusps, and can be traced out by a fixed point on the circumference of a circle of radius a which rolls on the outside of a fixed circle of radius $2a$, or by a fixed point on the

Figure 106

circumference of a circle of radius $3a$ rolling, with internal contact, on a fixed circle of radius $2a$.

The nephroid can be obtained by mathematical embroidery, as a line-envelope. Draw a circle, centre O, and let A be the end of a diameter. Starting at A, mark points on the circumference at intervals of 10°, and number them 0, 1, 2, 3, . . . the point A being marked 0. Now join the 1 to the 3, the 2 to the 6, the 3 to the 9, and in general *the point marked* n *to the point marked* 3n. The envelope of these lines is the nephroid (Figure 107).

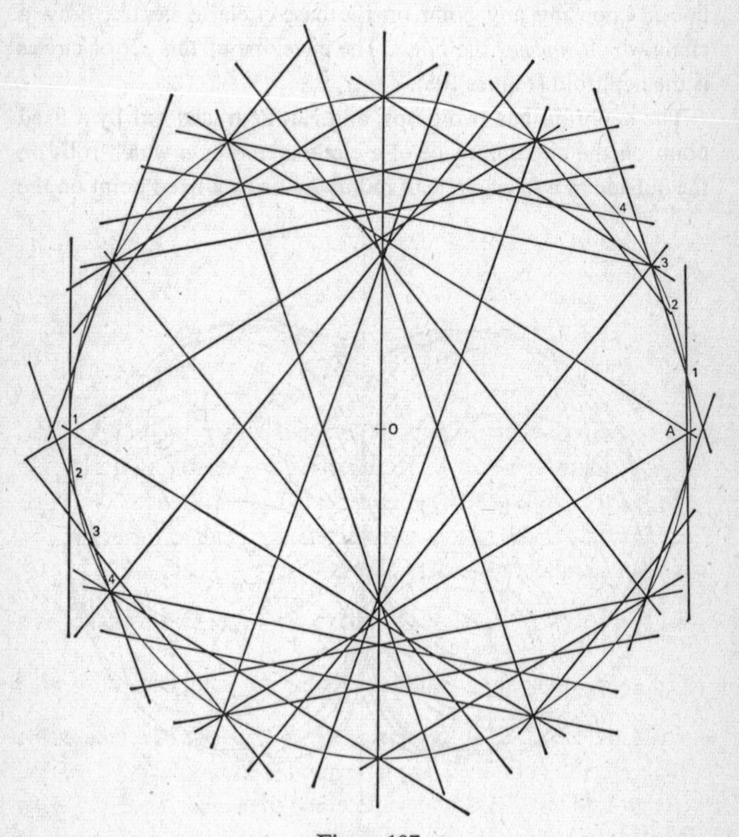

Figure 107

Envelopes of families of lines are important in optics, and were considered by Leonardo da Vinci. Suppose that we have a set of parallel rays falling on a polished circle, so that each ray is reflected at the point where it is incident with the circle (Figure 108).

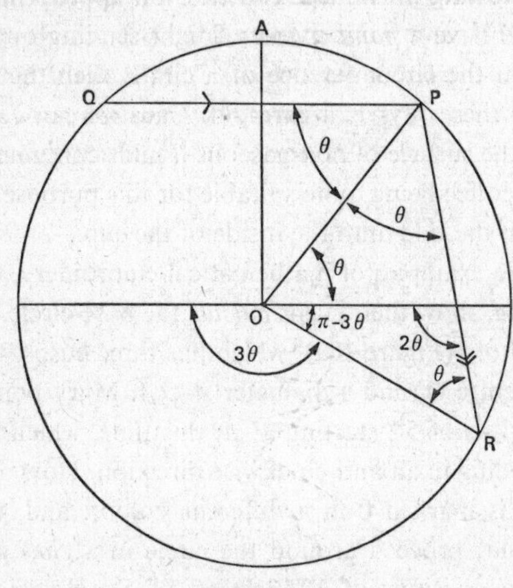

Figure 108

If O is the centre of the circle, QP the incident ray, the reflected ray is PR where the angle θ made by PQ with the radius OP is equal to the angle made by the reflected ray PR with OP. Then the envelope of the family of reflected rays, called a *caustic* of a circle for parallel rays, is a nephroid.

In Figure 108, $P\left(\frac{\pi}{2}-\theta\right)$ is joined to $R\left(\frac{\pi}{2}+\pi-3\theta\right)$ which equals $R\left(3(\frac{\pi}{2}-\theta)\right)$, indicating that the envelope is obtained by joining the point n on the circle to the point $3n$, so that the envelope is the same as the one obtained by mathematical embroidery.

According to E. H. Lockwood, the author of *A Book of Curves* (Cambridge University Press, 1961), the nephroid as a caustic may be observed by placing a dark-coloured cylindrical saucepan on the ground so that the rays from the sun, or from a powerful lamp, fall on it at an angle of about 60° to the horizontal. It should then be possible to see a good approximation to part of a nephroid.

Now that we have introduced caustics, it is appropriate to point out that if we have a *point source* of light sending out rays, the point lying on the circumference of a circle, then the caustic of the circle for these rays is a cardioid. Caustics can certainly be observed on the surface of non-gaseous liquids contained in cups, water, tea or coffee being quite suitable for this purpose. The light must fall from the side onto the inside of the cup.

Our last two examples of mathematical embroidery, the *deltoid* and the *astroid*, show their cusps *outside* the base-circle.

For the deltoid (Figure 109), which has three cusps, we draw a base-circle, centre O, and a diameter $A'OA$. Mark points on the circle at intervals of 5°, starting at A, this time, which is marked 0, and proceeding in an anti-clockwise direction. Now start again at A', which is marked 0 in a different colour, and, using this different colour, proceed around the circle in a *clockwise* direction, marking divisions of 10° instead of 5°. Now join points marked with *the same number in different colours.**

The name deltoid arises from the resemblance to the Greek capital letter Δ (delta). For the cusps to show up in mathematical embroidery, the straight line threads must be continued from the base-circle in both directions to points on the circumference of a larger, concentric circle. The threads inside the base-circle need not be visible. If the base-circle is of radius two inches, the larger circle should be at least six inches in radius.

The deltoid is a three-cusped hypocycloid, and can be obtained

*The highest number for the 10° division seems to be 35, whereas the highest number for the 5° division is 71. We keep on with the 10° numeration, so that 0 at A' is also marked 36, the point marked 1 is also marked 37, and so on. This enables us to use all the points on the 5° scale, and to obtain a complete figure.

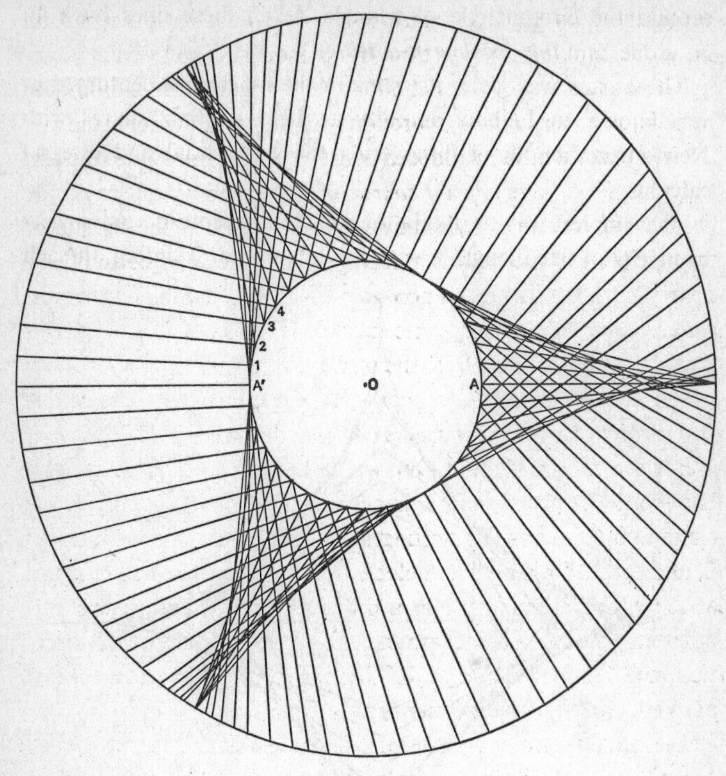

Figure 109

as a point locus, if a circle of radius a or $2a$ rolls on the inside of a fixed circle of radius $3a$.

It also arises in a fascinating way as an envelope of lines in connection with a fairly modern theorem in triangle geometry, called the *Simson Line Theorem.* If we take a point P on the circumcircle of a given triangle ABC, and drop perpendiculars PQ, PR and PS onto the sides BC, CA and AB respectively of the triangle ABC, it is not difficult to prove that the points Q, R and S always lie on a line, called the Simson line, after Robert Simson, a very influential eighteenth-century geometer. As P moves around the circumcircle of triangle ABC, these lines have an envelope, and *this envelope is a deltoid.*

The *astroid* was given its name in the nineteenth century, but was known to Leibniz as early as 1715. Leibniz shares with Newton the honour of discovering the differential and integral calculus.

The simplest way to obtain the astroid is to consider a line segment PQ of fixed length a with its ends P and Q sliding on each

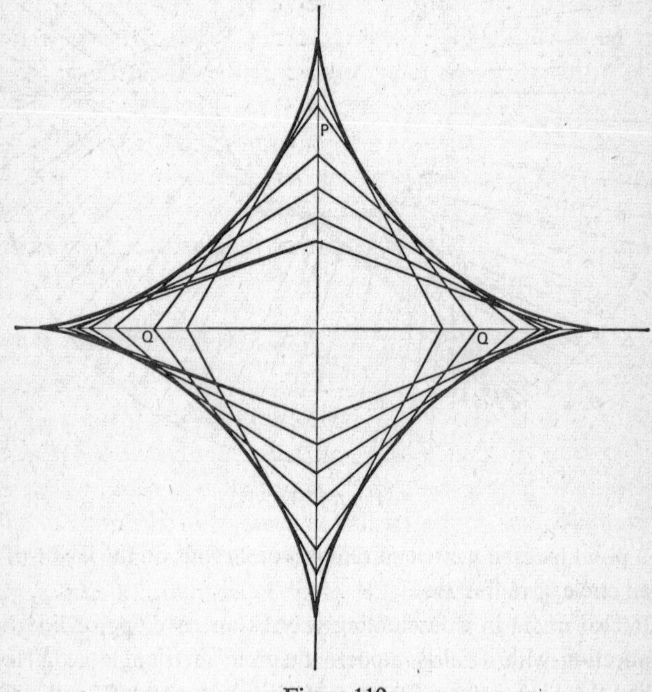

Figure 110

of two fixed lines at right angles to each other (Figure 110). If we fix the position of P at a distance from the intersection O which is less than a, we can find Q with compasses extended to the radius a. Joining PQ we obtain a family of lines which give the astroid as an envelope. The two lines are supposed to be continued through O, and the curve is symmetrical about each of the lines. The four cusps are at a distance a from O on each line.

To obtain the astroid by mathematical embroidery, we start with the usual base-circle, and a diameter $A'OA$, marking divisions of 5° from A, anti-clockwise, and numbering them 0, 1, 2, . . . We then begin again at A', and mark clockwise intervals of 15° with another colour, 0, 1, 2, . . . and so on. Joining points with the same number in different colours produces the astroid. The four cusps occur outside the base-circle, and the outer circle, to which the inner lines are produced, should have a diameter at least double that of the base-circle (Figure 111).

The astroid is a four-cusped hypocycloid, and can be produced as a point-locus by a point on the circumference of a circle of radius a rolling on the inside of a fixed circle of radius $4a$. The co-ordinates (x, y) of a variable point on the astroid can be written in terms of a variable t as:

$$x = 4a \cos^3 t, \qquad y = 4a \sin^3 t.$$

We cannot leave epicycloids and hypocycloids without considering the simplest of them all, the *cycloid*, which is the locus of a point on the circumference of a circle which *rolls along a fixed straight line*. This curve has an infinite number of cusps, traced as the point descends to the line and ascends again (Figure 112). It is a curve with some very remarkable properties, and engaged the attention of many mathematicians in the seventeenth century. It can be generated as an envelope, the envelope of a fixed diameter of a circle which rolls along a straight line.

If we turn the cycloid, as we have drawn it, upside down, and imagine it as a smooth curve in a vertical plane, a particle placed *anywhere* on it takes *the same time to reach the bottom*. This property inspired the great Dutch physicist and mathematician Christian Huyghens (1629–95) to use a cycloid around which a pendulum could wrap itself as it swung, hoping that the pendulum

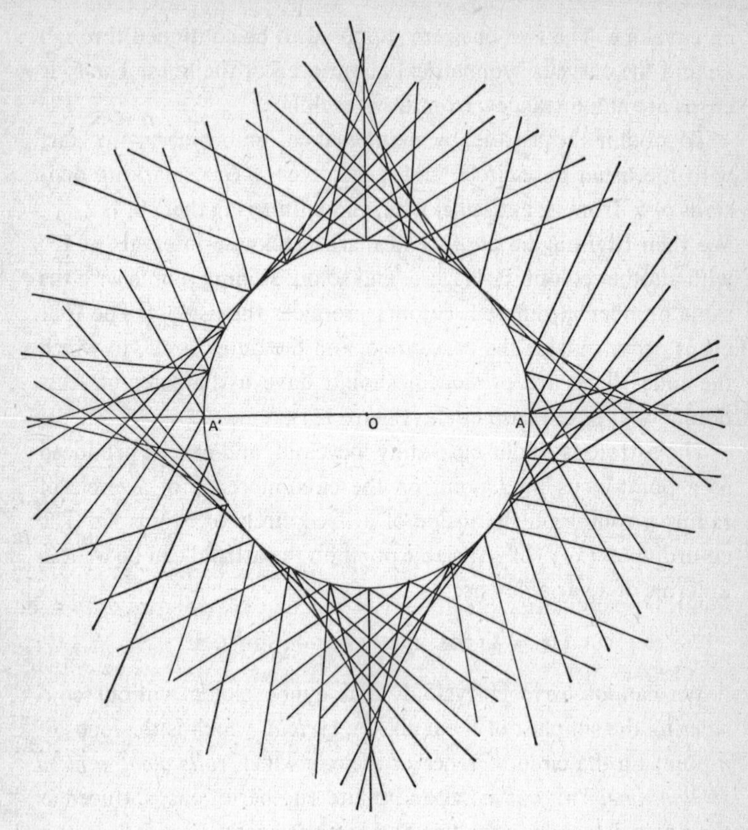

Figure 111

would become isochronous. If a string is held tightly up against a cycloid, and then unwrapped, any fixed point on the string describes a cycloid, and Huyghens thought that if the bob of the pendulum were constrained to move on a cycloidal arc, the ticks would be of equal duration. This, the first pendulum clock ever constructed, did not work very well, and was obsolete as soon as it was invented, but the idea is remarkably ingenious.

The final property of the cycloid we must mention is what was

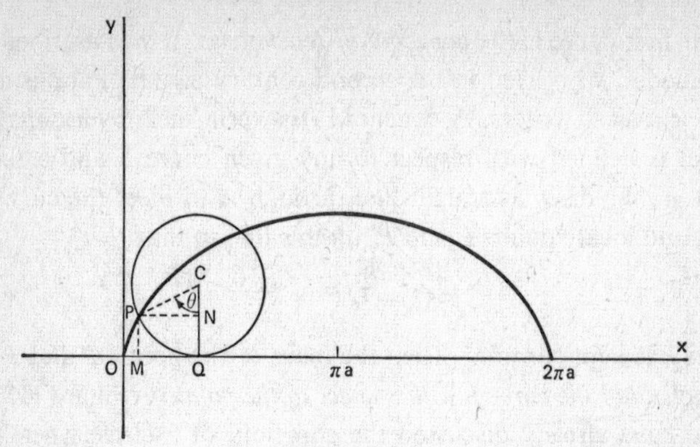

Figure 112

called the *Brachystochrone Property*, in the days when all mathematicians knew Greek. The cycloid is the curve down which a smooth particle will most quickly move under gravity from one given point to another. As far as I know, the escape chutes which airline companies sometimes use for a quick get-away from an aircraft are not cycloidal in shape. Of course *brachistos* means *shortest*, and *chronos* means *time*. The result is an unexpected one, and the proof is not at all simple.

The equations of a cycloid are most simply written in the form:

$$x = a(t - \sin t),$$
$$y = a(1 - \cos t),$$

where the rolling circle is of radius a, and t is a measure of the angle through which the circle has rolled, $t = 0$ being the starting point, when the moving point lies on the straight line, and the origin of co-ordinates is taken at that point.

There are, of course, many other beautiful curves which one would discuss in a book completely devoted to curves, but we now leave the fairly modern ones we have been investigating, and turn to some antique curves, invented to solve some outstanding problem in Greek geometry.

Our first curve is the *conchoid of Nicomedes*. It was ascribed to Nicomedes, who lived in the second century B.C. by Pappus and other classical writers. A conchoid (the term means *mussel-shell* shape) is defined with respect to any given curve S and a fixed point A. We draw a straight line through A to meet the curve S at Q, and locate points P and P' on this line so that

$$P'Q = QP = k,$$

where k is kept constant. Then the locus of the points P and P' is *a conchoid of the curve* S *with respect to the point* A (Figure 113).

We have already discussed the conchoid of a circle when the

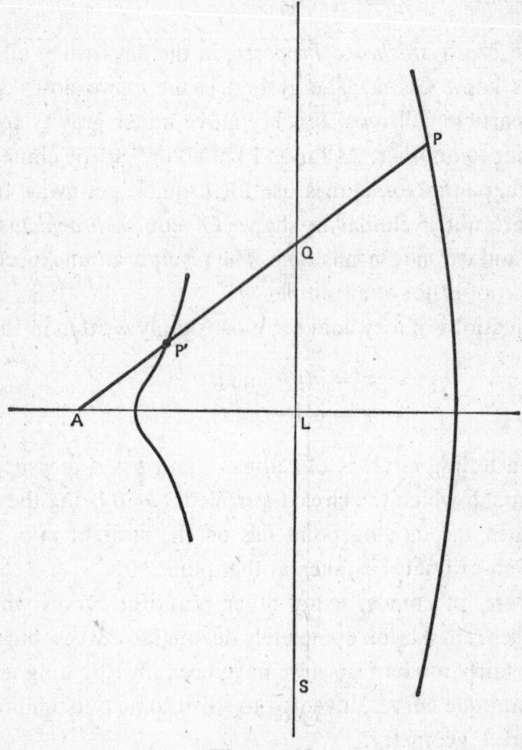

Figure 113

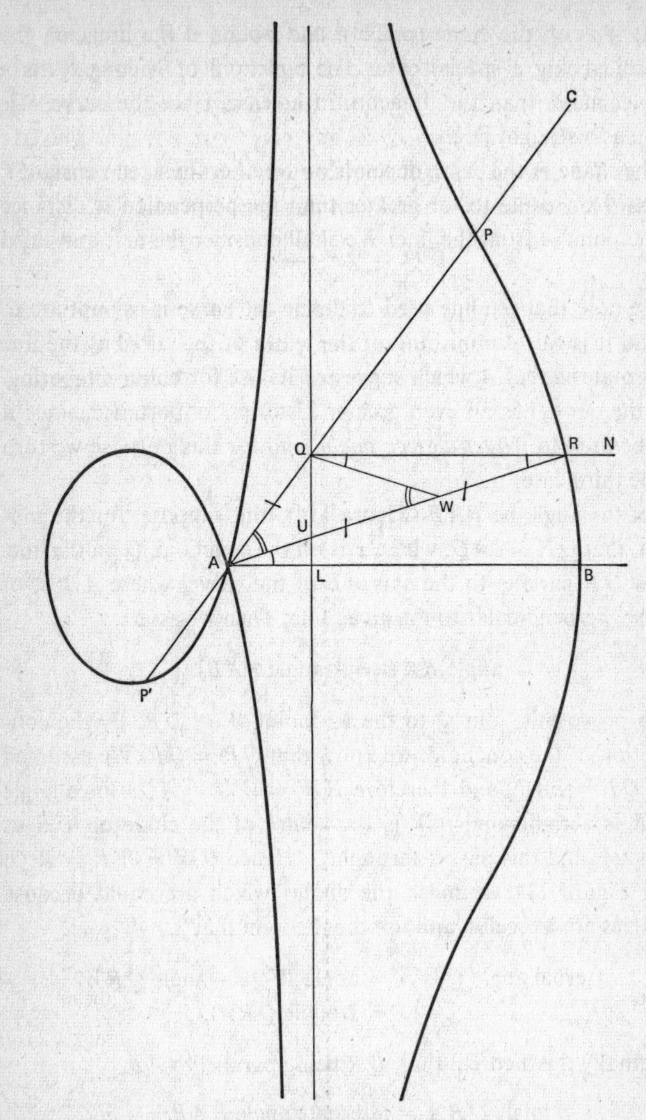

Figure 114

point A is on the circle (p. 229), and obtained the limaçon, the cardioid being a special case. The conchoid of Nicomedes is a simpler curve than the limaçon, in a sense, since the curve S is taken as a straight line.

The shape of the curve depends on whether the fixed constant k is less than, equal to, or greater than the perpendicular distance of the point A from the line. We shall consider the first and third case.

We note that the line used to define the curve is asymptotic to it, and it is this diminishing of the width of the curve as the line AQ rotates about A which suggested its use for column tapering. But the curve has an even greater historical importance, since it can be used to *trisect a given angle*, and for this purpose we turn to the third case.

Let the angle be PAB (Figure 114), and suppose, for the moment, that $QP = 2AQ$, where P is on the conchoid, Q on the line. Draw QR parallel to the axis AB of the curve, where AB is, of course, perpendicular to the given line. Then we assert:

$$\text{angle } RAB = \tfrac{1}{3}(\text{angle } PAB).$$

To prove this, join Q to the midpoint W of UR. By the construction of the conchoid, we know that $QP = UR$. We assumed that $QP = 2AQ$, and therefore $RW = WU = AQ$. Since angle UQR is a right angle, W is the centre of the circle on UR as diameter, and this passes through Q. Hence $UW = WR = WQ$. If in Figure 114 we mark the angles which are equal because triangles are isosceles, and use the theorem that

$$\begin{aligned}\text{external angle } QWA &= \text{angle } WQR + \text{angle } QRW\\ &= 2(\text{angle } QRA),\end{aligned}$$

and finally the theorem that, QR being parallel to AB,

$$\text{angle } QRA = \text{alternate angle } RAB,$$

we end up with

$$\text{angle } QAR = 2(\text{angle } RAB),$$

so that

$$\text{angle } RAB = \tfrac{1}{3}(\text{angle } PAB).$$

Now, how do we actually do this? We begin with the angle BAC, which we wish to trisect. Draw any line LQ at right angles to AB, cutting AC at Q. Draw QN parallel to AB. Now construct the conchoid of LQ with respect to the point A, with fixed distance $k = 2(AQ)$. Then if this conchoid cuts QN at R, we are back in the above situation, and the line AR trisects the angle BAC. Finally, we do not have to draw the conchoid! All that we have to do is to place a ruler so that its edge passes through A, and mark on the ruler the fixed distance $QP = 2(QA)$. Now slide the ruler with the edge always passing through A until the marks giving the fixed distance are on LQ and QN respectively. This gives us the points U and R, so the construction is a very speedy one.

Now, except for special angles, an angle cannot be trisected with the help of straight edge and compasses only. For example, 60° cannot be trisected. This can be proved, using algebraic methods developed within the past few centuries. Euclid was well aware of the impossibility of trisecting an angle, even if he was unable to prove the impossibility. But it will be asked, why is the method given above, using the marks on a straight edge, unacceptable?

The answer is that this method is not one of the Euclidean constructions, postulated on p. 142. We are allowed to join two points by a line, to draw a circle with given centre and radius, and, as a result of Euclid's Propositions, to place at a given point, as an extremity, a straight line equal to a given straight line.

Sliding a straight edge until one marked point on it falls on a given line, and another marked point on it falls on another given line is not a Euclidean construction. The essential distinction between Euclidean constructions and constructions such as the one we have just been discussing can only be revealed by algebra, and the tools of algebra which are powerful enough to deal with such problems were not developed until almost 2,000 years after Euclid's time.

The spiral of Archimedes can also be used for angle trisection, as we shall see when we discuss spirals (p. 248).

The *cissoid of Diocles* (cissoid means *ivy-shaped*) is mentioned by Geminus in the first century B.C., that is, about a century after the death of the inventor. We have a fixed point A on the circumference of a circle (Figure 115), and a straight line which touches the circle at the diametrically opposite point B. A straight line is drawn through A, cutting the circle again at Q and the line through B at R. A point P is taken on the line through A such that $AP = QR$, these lengths being measured in the direction indicated by the order of the letters. Then the locus of P is the cissoid of Diocles.

It is seen that the line through B is an asymptote to the curve, and that the curve also has a cusp at the point A. The equation of the curve, with axes along AB and at right angles to AB, is

$$y^2(2a-x) = x^3,$$

the radius of the circle being a.

We are interested in the cissoid because it was used to solve the problem of *duplicating the cube*. We can find a point U on OC such that $OU^3 = 2a^3$. A cubic altar of side OU will then have twice the volume of the original cubic altar of side a (p. 30).

To find U, we take the point L on OC where $OL = 2a$, and let BL meet the cissoid in P. We join AP, and then AP will meet OC in the required point U.

We prove this result, although once again the reader may omit the proof.

The equation of the line BL is $y = -2(x-2a) = 2(2a-x)$, and to find the point $P = (x', y')$ we substitute $2a-x = y/2$ in the equation of the cissoid, which gives us

$$(y')^3 = 2(x')^3,$$

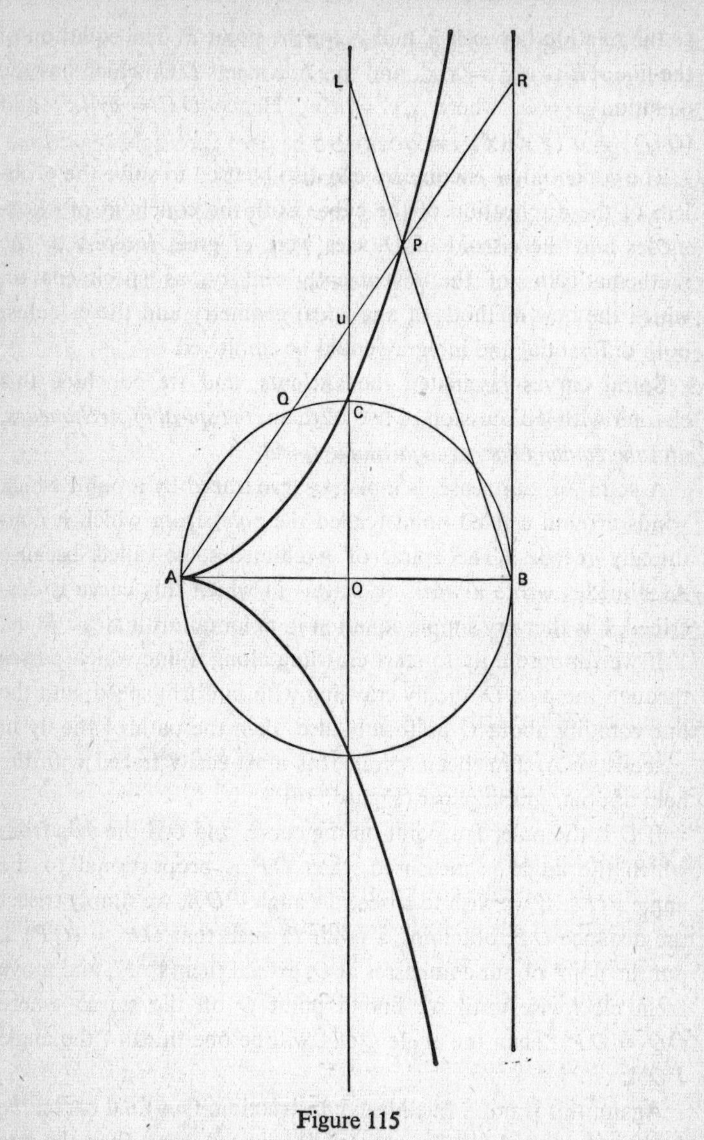

Figure 115

as the relation between x' and y' for the point P. The equation of the line AP is $y/y' = x/x'$, and this line meets OC, which has the equation $x = a$, where $y/y' = a/x'$. Hence $OU = ay'/x'$, and $(OU)^3 = a^3(y')^3/(x')^3 = 2a^3$.

The conchoid of Nicomedes can also be used to solve the problem of the duplication of the cube. Both the conchoid of Nicomedes and the cissoid of Diocles were of great interest to the mathematicians of the seventeenth century, as specimens on which the new methods of analytical geometry and the calculus, both differential and integral, could be employed.

Spiral curves fascinated the ancients, and we conclude this chapter with a discussion of two of them, *the spiral of Archimedes*, and *the equiangular*, or *logarithmic spiral*.

A spiral, in our sense, is a plane curve traced by a point which winds around a fixed point (called the *pole*) from which it continually recedes. The spiral of Archimedes, so-called because Archimedes wrote a work on spirals in which this curve is described, has the very simple equation in polar co-ordinates $r = a\theta$.

If we suppose a fly to start crawling along a line which passes through the pole O, the fly crawling with uniform speed, and the line rotating about O uniformly also, then the path of the fly in space is an Archimedean spiral. It is most easily traced with the help of polar graph paper (Figure 116).

If O is the pole, P a point on the curve, and OA the axis from which the angle is measured, then OP is proportional to the angle POA. If we wish to trisect the angle POA, we simply trisect the distance OP, obtaining a point P' such that $OP' = (OP)/3$, put the point of our compasses at O, extend them to P', and move them clockwise until we find a point Q on the spiral, where $OQ = OP'$. Then the angle QOA will be one third of the angle POA.

Again, this is not a Euclidean construction. Our final curve, *the equiangular spiral*, is a far more sophisticated curve than the one considered by Archimedes. It was first studied by Descartes, in

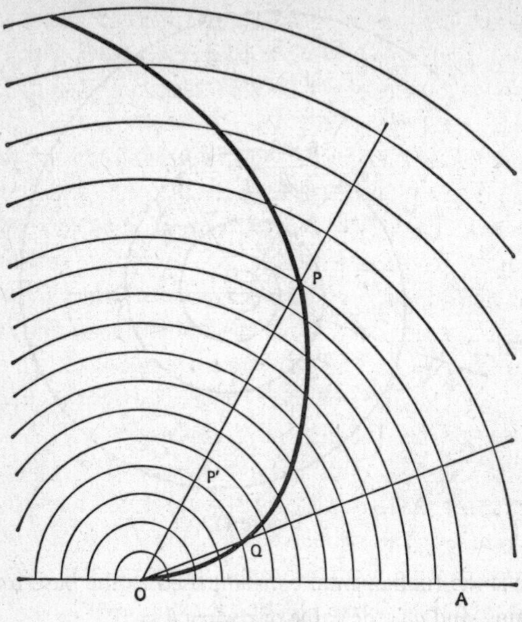

Figure 116

1638, Torricelli worked on it independently, and towards the end of the seventeenth century John Bernouilli found many of its remarkable properties, which we shall discuss. He was so impressed by these almost mystic properties, that he asked to have the curve cut on his tombstone, with the words: *Eadem mutata resurgo* – Though changed I rise unchanged.

The equiangular spiral (Figure 117) can be defined as the locus of a point P which moves in a plane so that the tangent at P makes a *fixed angle* α with the outward-drawn radius vector OP, the point O being fixed. By the procedures of the differential calculus, the equation of the curve is quite easily obtained, the natural co-ordinates being polar (r, θ)-co-ordinates, and it has the attractive form:

$$r = ae^{\theta \cot \alpha}$$

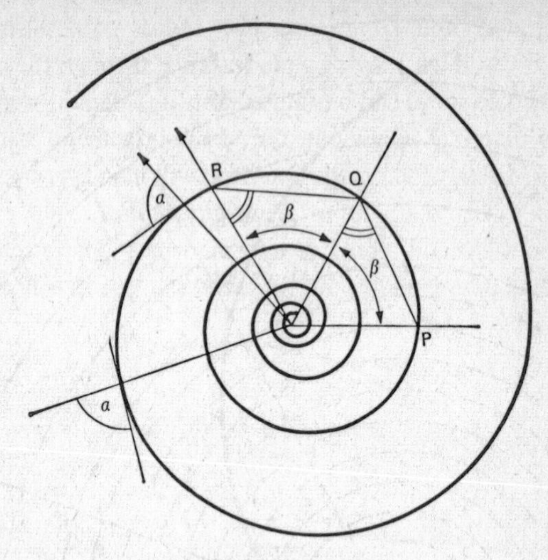

Figure 117

where e is the fundamental constant used as the base for natural logarithms, and a is the value of r when $\theta = 0$.

If we use one of the fundamental properties of what is called *the exponential function* e^x, which says that

$$e^{p+q} = e^p . e^q,$$

we can obtain a fundamental property of the equiangular spiral.

Suppose that P, Q, R, ... are points on the curve which are *equally spaced*, so that the angles $QOP, ROQ, \ldots$ are all equal to β, say. Then

$$OQ/OP = e^{\beta \cot \alpha} = OR/OQ = \ldots$$

and so in the triangles $OPQ, OQR, \ldots$ the angles at O are equal, and the ratio of corresponding sides is the same. Hence, by a Proposition of Euclid, the triangles are all similar to each other, which means that the angles OQP, $ORQ, \ldots$ are all equal. From this property alone we can show that the tangent at any point P of the curve makes a constant angle with OP.

Again, if P and Q are any two points of the spiral, with $OP > OQ$, and we imagine a pin fastened through the pole O, so that the whole spiral can be rotated about O, then if we rotate the spiral so that OQ falls along OP, an enlargement of the spiral, keeping O fixed, which maps Q onto P will map every point of the rotated spiral onto a point of the original spiral.

Certain shells and fossils have forms which closely resemble the equiangular spiral, and Figure 118 shows a section of the shell of a

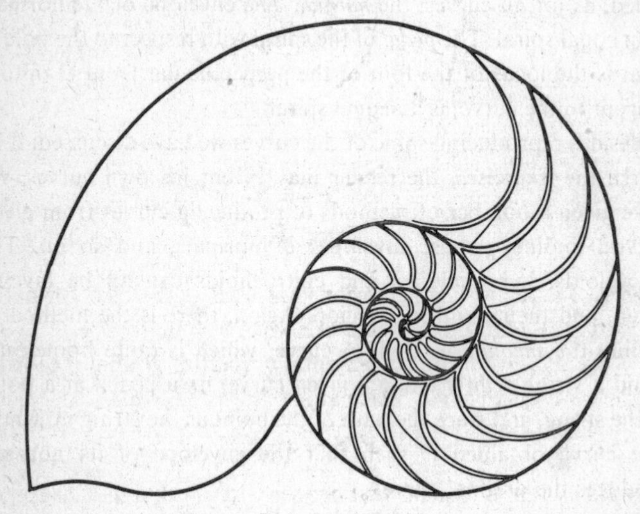

Figure 118

Pearly Nautilus fossil. Books on growth and form have theories as to why this is so. There is a very well-known one by D'Arcy W. Thompson, *On Growth and Form.*

In Chapter 4, p. 116 we described an artificial spiral constructed of quadrants of circles, and associated with the continued division of rectangles whose sides are in the golden number into squares and further golden rectangles. There is an equiangular spiral which closely approximates this spiral, but the true spiral cuts the sides of the successive squares at very small angles, instead of touching them. Of course, it is a matter of taste as to which spiral is called the true, and which the artificial.

We end this chapter by giving a few of the properties which intrigued John Bernouilli so profoundly.

If a ray of light from a source at O is reflected by the curve at P, the envelope of the reflected ray, as P varies, is an equal spiral. This means that the curve has a *caustic* of the same shape. At each point of the curve there is a perpendicular to the tangent, called, as for all curves, the *normal*. The envelope of the normals is an equal spiral. The *pedal* of the spiral with respect to the pole O (that is the locus of the foot of the perpendicular from O onto a tangent to the curve) is an equal spiral.

Besides reproducing some of the curves we have discussed, if he works the Exercises, the reader may invent his own curves. We have given a number of methods of producing curves from given curves, finding pedals, envelopes of normals, and so on. The epicycloids, hypocycloids, and epitrochoids can all be investigated, and their number is legion. Again, there is the method of finding the *involute* of a given curve, which is quite homespun. Wind a string tightly along a given curve, fix a pencil at a point of the string, and trace the path of the point as the string unwinds. The curve obtained is such that the envelope of its normals produces the original curve!

Many of the envelopes we have described have attracted the attention of artists fairly recently, and are offered for sale at high prices, the work, carried out with coloured threads on a fine piece of board, offering an opportunity for transcendental meditation. Mathematical embroidery was suggested for schoolgirls almost a century ago, and it is bound to become commercialized. A recent apparatus for drawing epicycles has had an enormous success. Curves will always, one can be certain, remain one of the most fascinating creations of mathematics.

Exercises

1. Draw about 20 lines touching a parabola, by the method given on p. 202 (S is a fixed point, l a fixed line, P any point on l, and the line PR is drawn perpendicular to SP. The lines PR touch a parabola).

2. Obtain a set of lines touching a parabola by paper creasing. (The line m is drawn on a sheet of slightly waxed paper, and a point S not on the line is marked. Fold the paper so that m passes through S, and crease it neatly: that is, crease the paper along a straight line so that m passes through S. Then crease the paper along another fold so that again m passes through S and so on.)

Can you reconcile this method with the one suggested in Exercise 1 above? (If Q is the point on m which, after folding, lies on S, and SQ meets the crease in P, observe that the points P lie on a line, and the angle between SP and the crease is a right angle.)

3. Use the parabola obtained from either of the preceding Exercises to verify the theorem that rays parallel to the axis are reflected at the curve so as to pass through a fixed point, the focus S of the parabola.

4. Draw two lines l and m intersecting at a point O, and mark off equal divisions on l, beginning at O, numbering them, O being marked as nought. Do the same on m, the scale not necessarily being the same. Join the point marked 10 on l to the point marked 1 on m, the point marked 9 on l to the point marked 2 on m, the point marked 8 on l to the point marked 3 on m, and so on. These lines touch a parabola also, and l and m are tangents to this parabola.

(This is the method used for producing large-scale mathematical embroidery, since the arcs can be fitted all over a sheet of paper. The theorem on which the construction is based says that if a variable tangent to a parabola cuts two given tangents l and m in the points P, Q, R, ... and P', Q', R', ... respectively, then the ratios $PQ:QR:\ldots$ and the corresponding ratios $P'Q':Q'R':\ldots$ are equal.)

5. Obtain tangents to an ellipse by paper-creasing. (Draw a circle, and let S be a fixed point inside the circle. Fold the paper so that after creasing along a line, a point of the circle lies on S. The creases will touch an ellipse, with one focus at S.)

6. Can you obtain the tangents to a hyperbola by paper-creasing?

7. Construct an ellipse, given as a plane section of a right circular

cone, by Dürer's method (p. 214). See if you can be sufficiently accurate to avoid making the curve egg-shaped!

8. Draw a cardioid by the method described on p. 221. (Join a fixed point A on a circle of suitable diameter a to a point Q on the circle, and extend AQ by a distance a to P. The path traced by the points P as Q moves around the circle is a cardioid.)

9. Obtain a cardioid as an envelope of circles (p. 220). (Let A be a fixed point on a circle, P any point on the circle, and draw a circle centre P and radius PA.)

10. Obtain a cardioid as an envelope of straight lines, using either coloured threads or pencil lines. (Let $A'OA$ be a diameter of a circle [p. 224]. Starting at A, mark divisions of 10° on the outside of the circle, A being 0. Now begin at A', and on the inside of the circle, with the same sense of rotation, mark intervals of 20° as the points 1, 2, . . . the point A' being marked 0. Now join points on the outside of the circle to points on the inside with the same number.)

11. Obtain a deltoid as an envelope by mathematical embroidery. (Let $A'OA$ be a diameter of a suitable circle, and starting at A, which is marked 0, mark off points at intervals of 5° in an anti-clockwise direction. Start again at A', and mark off points at intervals of 10° in a clockwise direction, and in another colour. Join points with the same number and with different colours. The cusps lie outside the circle [Figure 109], so the threads must be extended to points on the circumference of a larger, concentric circle, of at least three times the radius of the original circle. The apparent impossibility of joining *coincident* points, such as the points marked 12 in different colours, is overcome by drawing the tangent to the circle at the point, and this is the required join.)

12. Invent your own curves, using mathematical embroidery, or any other systematic method.

13. Those who carry out Exercise 2 in the set following Chapter 3 become curious as to the ellipse traced out by the vertex C of a triangular cut-out ABC when A moves on a given line l, and B moves on a given line m. Let O be the intersection of l and m. Draw a perpendicular to l at A, and a perpendicular to m at B, and let these perpendiculars intersect at I. Draw the circle on OI as diameter, and label it c. Draw the circle centre O which passes through I, and label it d. It can be

shown that I is *the instantaneous centre* for the motion of triangle ABC, which means that instantaneously the triangle is rotating about I. But in this case the finite motion of triangle ABC is given by the rolling of circle c on the fixed circle d.

Join C to the centre of c, and let this diameter of c meet it in the points P and Q. Then OP, and OQ, which are perpendicular, are the principal axes of the ellipse described by the point C, and C traces the same ellipse as if C were a fixed point on the line segment of fixed length PQ, where P moves on OP and Q moves on OQ. Verify that points on the ellipse constructed one way are also points on the ellipse constructed the other way.

This analysis shows that if the triangle ABC is such that C is at Q, say, then the path of C is the line OQ. If C is at P, then the path is the line OP.

·9·

SPACE

In this final chapter we consider *space* from a number of aspects. Of course, we have been doing this all through the preceding chapters, but here we can bring together a number of ideas which have already come up, and discuss them more systematically.

Regular polygons have appeared a number of times, and we have tacitly assumed that they were *convex*. By this we mean that

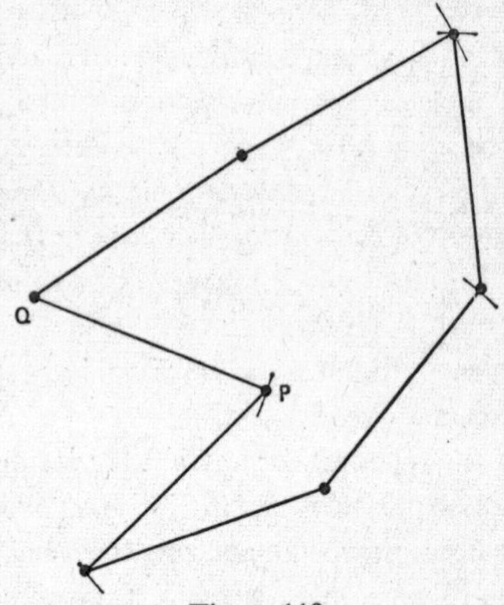

Figure 119

if P and Q are two points which define any side, then the polygon lies entirely on one side of PQ. Hence the kind of polygon we see in Figure 119 is not considered.

Our regular polygons have equal sides and angles, and their

vertices lie on a circle, and there exists another circle which touches all the sides. It is therefore clear that the construction of a regular polygon with n vertices and n sides is equivalent to the division of the total circumference of a circle into n equal parts. If we solve the one problem, we have solved the other.

From Euclid we know how to construct a right angle, and how to bisect any given angle, so that we can form half a right angle, a quarter of a right angle, and so on. Since the construction of a right angle is equivalent to the construction of a regular polygon of $n = 4$ sides, we can also construct regular polygons of 4(2) sides, of $4(2)^2$ sides, . . . and eventually of $4(2)^n$ sides, for any integer n.

We can construct an equilateral triangle, which is a polygon of the type we are investigating with $n = 3$ sides, each side subtending 120° at the centre of the circumscribing circle, and by continual bisection we can obtain a figure with $3(2)^n$ sides. Again, we saw how to construct a regular pentagon (p. 67), and so we can obtain a regular polygon, by angle bisection, with $5(2)^n$ sides.

Finally, since $24 = 60-36$, and $36 = 72/2$, and we can construct angles of 60° and 72°, we can construct a polygon with sides which subtend 24° at the centre of the circumscribing circle, and this is a regular polygon with 15 sides. By bisection of angles we can now construct a polygon with $15(2)^n$ sides.

These constructions must all be Euclidean, of course, with ruler and compasses only.

It might be thought that for any n it is possible to construct a regular polygon with n sides, but in 1796 Karl Friedrich Gauss, at the age of nineteen, proved an amazing theorem. This says that a regular polygon with n sides, *where* n *is prime*, can only be constructed with ruler and compasses if

$$n = 2^{2^m}+1.$$

The index of 2 is 2^m, and we take $2^0 = 1$, so that when $m = 0$, $n = 3$, when $m = 1$, $2^2 = 4$ and $n = 5$, when $m = 2$, $n = 17$, when $m = 3$, $n = 257$, and when $m = 4$, then $n = 65,537$. These numbers are all *prime**, without factors, and therefore, by the theorem above, they can be constructed.

It would have delighted the Greeks to know this theorem, which is proved algebraically. Its discovery helped young Gauss to make his mind up to do mathematics instead of law! As we saw with the regular pentagon, the construction of regular polygons is by no means obvious, and the algebraic proof of the possibility of the construction of regular polygons with 257 and 65,537 sides does not help too much with the details of construction. In fact, some enthusiasts have devoted their whole adult lives to the last construction. But at least, they were not chasing a chimera. Angle trisectors, as we have already remarked, are engaged on a hopeless mission, but perhaps this appeals to some people.

When discussing Leonardo da Vinci's geometrical discoveries, we promised to return to the notion of the *symmetries* of a geometrical figure (p. 97). Leonardo investigated the symmetries of a regular octagon. For simplicity, we begin with a square.

Imagine a cardboard square laid on a plane with two perpendicular axes, with the centre of the square on the origin of co-ordinates, and one side of the square horizontal (Figure 120), the other vertical, so that the plane itself is a vertical plane. We now ask what rigid motions, keeping the centre O fixed, will keep the square, as a geometrical figure, unchanged?

The square has rotational symmetry. It is carried into itself by the following rotations, which we give names:

R: a rotation clockwise through 90° around the centre O.
R' and R'': similar rotations through 180° and 270°.

*Fermat (1601–65) conjectured that all integers of the form indicated are prime, but Gauss showed that when $m = 5$, $n = 2^p+1$, where $p = 64$, is composite. For further information see p. 14 of *An Introduction to the Theory of Numbers* (G. H. Hardy and E. M. Wright, Oxford, Clarendon Press, 1960).

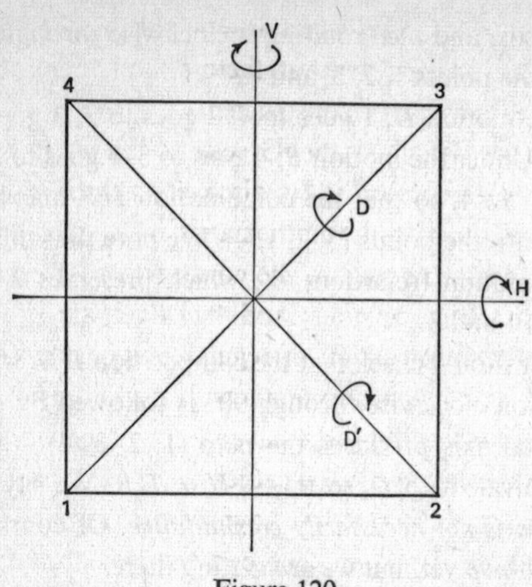

Figure 120

The square also has *reflective* symmetry. It can be carried into itself by the following rigid reflexions:

> H: a reflexion in the horizontal axis through O.
> V: a reflexion in the vertical axis through O.
> D: a reflexion in the diagonal marked 1 3 in Figure 120.
> D': a reflexion in the diagonal marked 2 4 in Figure 120.

This gives us seven symmetries so far. In case the reader is confused by the definition of symmetry given on p. 18 and the procedure here, we point out that we are not merely considering one-to-one transformations which map the four vertices of the square onto four vertices of the square, but only those which preserve the *structure* of the square as a geometrical figure. This means that our transformations are *isometries*, maps which *preserve* the *distance* between two points mapped.

The real advance from the fairly obvious concepts we have described lies in the notion of the *composition* of rigid motions. We define HR to be the result of first reflecting the square in the horizontal axis, and *then* rotating it clockwise through 90°. What happens to the points 1, 2, 3, and 4?

Under the motion H, 1 goes to 4, 2 goes to 3, 3 goes to 2, and 4 goes to 1. Under the motion R, 4 goes to 3, 3 goes to 2, 2 goes to 1 and 1 goes to 4, so that the combination HR maps the points (1, 2, 3, 4) onto the points (3, 2, 1, 4). We note that this is equivalent to the motion (reflexion) D', which preserves 2 and 4 and interchanges 1 and 3.

The reader should check that the composition RH, which means that a rotation clockwise through 90° is followed by reflexion in the horizontal axis produces the map $(1, 2, 3, 4) \rightarrow (1, 4, 3, 2)$ which is equivalent to D, so that $RH \neq HR$. We say that *group multiplication is not necessarily commutative.* Of course, we have not defined *group* yet, but we are getting there.

If we take the composition of either D or D' with itself, the square returns to its original position. It is extremely useful to say that this mapping, which changes nothing, is also one of the symmetries of the square. It is called the *identity*, and labelled I, and raises the number of symmetries of the square to eight.

These eight symmetries form what is called a *closed set* under composition of symmetries. Any composition of these eight symmetries produces a symmetry of the set.

Again, any symmetry of the set has an *inverse*, which means that associated with any symmetry, say T, of the set, there is another symmetry T' of the set, which may coincide with T, which is such that $TT' = I$.

In this case D and D' are their own inverses. Finally, if we take any three of our symmetries, which we call T, T' and T'', $T(T'T'') = (TT')T''$, which means that if we evaluate $T'T''$, and then compose it with T, the resulting symmetry is the same as if we first compose T and T', and then compose this with T''. We say that composition is *associative* in such a case.

We have now shown that the symmetries of a square obey the axioms for a *group* of transformations. The notion of a group is of the utmost importance in mathematics, and we can only refer the reader to a few of the many books on the theory of groups if he is interested in pursuing the subject further.

Our present group of transformations can be verified to consist of the following eight symmetries: I, R, R^2, R^3, H, HR, HR^2, HR^3, and the reader should verify this list. It is evident that R composed with itself gives R', a rotation through 180°, and we saw that $RH = D$, and $HR = D'$. Although RH does not occur in the list, it is easy to verify that $RH = HR^3$.

The group can be said to be *generated* by the symmetries R and H, since all elements of the group arise from the composition of just these two symmetries. According to Herman Weyl*, Leonardo found the generators of the group of symmetries of a regular n-gon, although, as we remarked, Leonardo's Notebooks only reveal the case $n = 8$.

Groups of symmetries arise in the theory of ornaments along a line, in the theory of two-dimensional patterns, in the theory of crystals, in the theory of packing, and in many other branches of mathematics. For some of these topics we refer to *The Gentle Art of Mathematics*†, and for a more general survey to *Symmetry*, by Hermann Weyl.

In three dimensions the fundamental completely regular surfaces are the so-called Platonic solids. These are mentioned by Plato in the *Timaeus*, but were certainly known to the Pythagoreans, centuries earlier. For any such surface each face is congruent to every other face, and the surface lies completely on one side of any face. The surfaces are shown in Figure 25, and their faces are either squares, triangles or pentagons.

In the *Timaeus* these *regular polyhedra*, as they are also called, are constructed out of right-angled triangles. Plato picks out two as being fundamental, the isosceles right-angled triangle, which is

**Symmetry*, Princeton University Press, 1952, p. 66.

†Pedoe, English University Press, 1965, p. 104.

half of a square, and the triangle with angles 30°, 60° and 90°, obtained by bisecting an equilateral triangle.

'The reason,' says Plato, 'is too long a story: but if anyone should challenge us and discover that it is not so, he is more than welcome to the prize.'

Out of four equilateral triangles Plato constructs a regular tetrahedron. This is the simplest, and so the most fundamental of the regular solids, and Plato associates it with *fire*, the most fundamental of the elements.

With eight equilateral triangles he constructs a regular octahedron, which is assigned to *air*, and twenty equilateral triangles produces the regular icosahedron, which is assigned to *water*.

These are the only solids made out of equilateral triangles, or, if you will, halves of equilateral triangles. Two of Plato's isosceles triangles, reconstituted, make a square, and six squares put together give a cube, a hard resistant body assigned to *earth*.

There remains the dodecahedron, each side of which is a regular pentagon, but Plato dismisses this casually with the sentence: 'There still remained one construction, the fifth, and the god used it for the whole, adorning it with a pattern of animal figures.' There are different translations of Plato, of course, and the one I have used explains that the animal figures probably referred to the signs of the zodiac and other constellations in the sky.

There are equations which connect the number of faces f, of edges e and of vertices v of a given regular convex polyhedron. The first equation, due to Euler (1707–83), is

$$f+v = e+2.$$

If there are n sides in any face, and m edges emerge from any vertex,

$$nf = mv = 2e,$$

which is obtained by counting. Five solutions of these equations are given in the table below, and it is not difficult to show that *no others exist*. The solutions correspond to the Platonic solids.

Solid	*f*	*v*	*e*	*m*	*n*	*Face*
Tetrahedron	4	4	6	3	3	Triangle
Hexahedron (cube)	6	8	12	3	4	Square
Octahedron	8	6	12	4	3	Triangle
Dodecahedron	12	20	30	3	5	Pentagon
Icosahedron	20	12	30	5	3	Triangle

The reader should make a set of the Platonic solids, and they can be made quite easily by sticking triangles, squares or pentagons together. He will then discover why there is no Platonic solid with a hexagon (6-gon) as a face. When three pentagons are stuck together, so that there is a vertex with three edges emerging from the vertex, the sum of the angles as we go around the vertex is $3(108) = 324°$. This is less than 360°. If we take three hexagons and stick them together, the sum of the angles at the vertex is $3(120) = 360°$, and the three hexagons *lie in a plane*. In order to form a *solid* angle, in which the faces meeting at the vertex of the angle do not lie in a plane, the sum of the angles at the vertex must be less than 360°.

The method of forming solid angles is described by Plato. For example, to construct the icosahedron, which consists in all of twenty equilateral triangles, start by constructing solid angles with five equilateral triangles clustering round a vertex. It will then be seen how to fit the solid angles together.

In the *Divina Proportione* of Luca Pacioli, one edition of which appeared in Venice in 1509, there are careful perspective drawings of models of the Platonic solids. The models were made by Pacioli, and often given as presents to eminent personages, and the drawings are attributed to Leonardo da Vinci, who was a friend of this remarkable monk.

Pacioli was a Franciscan, and an enthusiastic mathematician. He taught at various universities in Italy, and must have been a great teacher. He published other books besides the *Divina Pro-*

portione. In one he taught the method of double-entry in book-keeping, the laws of arithmetic, and how to run a business successfully. This *Summa de Arithmetica, Geometria, Proportioni e Proportionalità*, published in 1494, had an enormous influence. Problems, all too familiar, about taps filling baths with holes in them, can be found in the *Summa*, but they originated elsewhere. The *Summa* is in eight parts, *à reverentia de la 8 beatitudine*. It is also believed that Pacioli published a translation in the vernacular (Italian) of Euclid, but this has been completely lost. His translation of Euclid into Latin appeared in 1509. His style in the *Divina Proportione* is inimitable. The book is dedicated to the Duke of Milan, Ludovico Sforza, with many compliments to the Duke spread all through the text. There are thirteen descriptions of properties of the divine ratio, (our *golden* ratio), the first a mere affirmation that the theorems of Euclid are assumed to be known. Then come the following attributes to the various properties of the golden ratio. Pacioli writes:

> – del secondo *essentiale* effecto de questa proportione, del terzo suo *singolare* effecto, del quarto *ineffabile* effecto, del quinto suo *mirabile* effecto, del suo sexto *innominabile* effecto, del septimo suo *inextimabile* effecto, del decimo suo *supremo* effecto, del suo undecimo *excellentissimo* effecto, del suo duodecimo *incomprehensibile* effecto, del terzodecimo suo *dignissimo* effecto . . .

Pacioli then pauses for breath, and remarks that there is not enough ink and paper in existence to give all the properties of the divine ratio, and to stop at thirteen does honour to *the twelve disciples and their holy Head, Jesus Christ*. He then congratulates Sforza again for his piety, and for getting Leonardo da Vinci to paint *The Last Supper* in the Church of Santa Maria delle Grazie.

We discussed the divine ratio in Chapter 4. Many of the admirable properties admired by Pacioli are exhibited by the successive ratios shown in Figure 41.

There is a very attractive painting (Plate 24), by Jacopo de

Barbari, of Fra Luca Pacioli and his student Guidobaldo, Duke of Urbino. We note a dodecahedron on the table, and the display of another solid, called a *semi-regular Archimedean solid.*

The semi-regular polyhedra are those which admit a variety of polygons as faces, provided that they are all regular, and that the vertices are indistinguishable, the one from another (Figure 121).

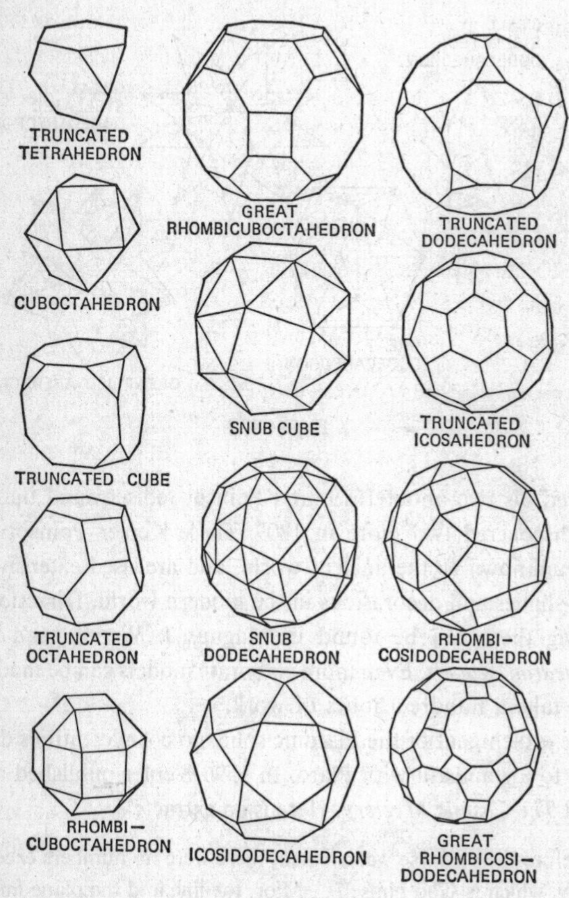

Figure 121

Plato is said to have known at least one, the cuboctahedron, and Archimedes wrote about the entire set, though his book on them is lost. Dürer gives the *nets* for some Archimedean solids in his *Underweysung* (p. 76), but they were first treated systematically by Kepler, who also constructed two of the *non-convex* solids called the *Kepler-Poinsot solids* (Figure 122). Kepler dis-

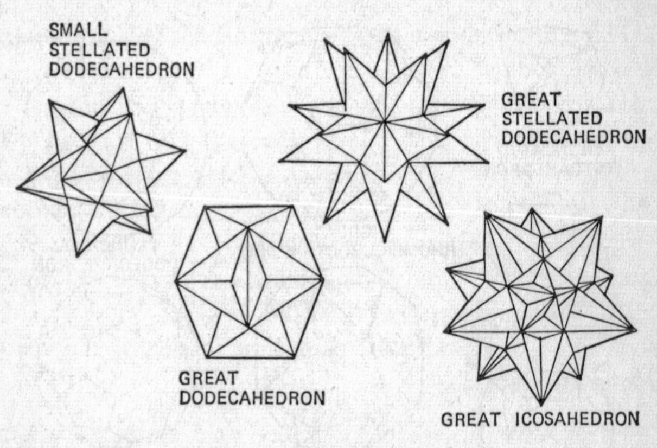

Figure 122

covered his two about 1619, and Poinsot rediscovered these two and discovered two more in 1809. These Kepler–Poinsot solids were unknown to the ancient world, and are used extensively as lamp-shades and decorations in the modern world. Directions for making them can be found in Magnus J. Wenninger's book: *Polyhedron Models*. Even more elaborate models can be made, but some take a hundred hours of work.

The attachment of the Platonic solids to cosmic entities did not come to a standstill with Plato. In 1596 Kepler published a tract called *The Cosmic Mystery*. Here is an extract:

. . . Before the universe was created, there were no numbers except the Trinity, which is God himself . . . For, the line and the plane imply no numbers: here infinitude itself reigns. Let us consider, therefore, the solids. We must first eliminate the irregular solids, because we are only concerned with orderly creation. There remain six bodies, the sphere and the five regular polyhedra. To the sphere corresponds the outer heaven. On the other hand, the dynamic world is represented by the flat-faced solids. Of these there are five: when viewed as boundaries, however, these five determine six distinct things: hence the six planets that revolve about the sun. This is also the reason why there are but six planets . . .

. . . I have further shown that the regular solids fall into two groups: three in one, and two in the other. To the larger group belongs, first of all, the Cube, then the Pyramid, and finally the Dodecahedron. To the second group belongs, first, the Octahedron, and second, the Icosahedron. That is why the most important portion of the universe, the Earth – where God's image is reflected in man – separates the two groups. For, as I have proved next, the solids of the first group must lie beyond the earth's orbit, and those of the second group within . . . Thus I was led to assign the Cube to Saturn, the Tetrahedron to Jupiter, the Dodecahedron to Mars, the Icosahedron to Venus, and the Octahedron to Mercury . . .

To emphasize his theory, Kepler envisaged an impressive model of the universe which shows a cube, with a tetrahedron inscribed in it, a dodecahedron inscribed in the tetrahedron, an icosahedron inscribed in the dodecahedron, and finally an octahedron inscribed in the dodecahedron.

When this essay reached Galileo and Tycho Brahe, both astronomers replied with flattering comments, and Tycho Brahe was so impressed that he invited the young Kepler to become his assistant, urging him at the same time to apply his methods to the Tychonian cosmic system, which was a sort of cross between the Ptolemaic and the Copernican, the planets spinning round the sun while the sun executes a pirouette about the earth.

Of course, the planets Neptune and Uranus were not known in Kepler's time, and one wonders how Kepler would have modified his theory if they had been discovered during his lifetime. Whatever his private view of the universe, the fact is that Kepler did discover the three fundamental laws of planetary motion, using Tycho Brahe's observations:

I. The planets move in ellipses with the sun at a focus.

II. In the motion of a planet, sectors of equal area are swept out by focal radii in equal times.

III. The square of the period of a complete revolution of a planet about the sun is proportional to the cube of the major axis of the orbit.

As we have already indicated, Newton, in *Principia Mathematica* a century later, was able to show, using geometry only, that these three laws can be deduced mathematically from the inverse square law of gravitation. Newton arrived at some of his results by other methods, but felt that they were 'too harsh' for public view, and converted them into geometrical methods.

Newton himself regarded his life-work as his explanations of the prophecies of the prophet Daniel. Assuming that every prophecy in the Book of Daniel is true, Newton was able to prove, to his own satisfaction, that the prophecies had been confirmed.

There are modern mystics, of course. When I was professor of mathematics in the Sudan, I was visited one morning by a Greek merchant, who waved a detailed diagram of the Great Pyramid of Gizeh which, he asserted, suitably interpreted, held the key to everything which had ever happened and was going to happen. On the day that the Second World War broke out, I listened to an orator on Southampton Common explaining how the British Israelites had foreseen the war, again from the Great Pyramid of Gizeh. I forget how far ahead in time the prophecies reached.

Some modern mystics know a fair amount of mathematics. Peter Demianovich Ouspensky 'combines the logic of a mathematician with the vision of a mystic in his quest for solutions to the problems of Man and the Universe'. He was born in Moscow in 1878, and, again according to the blurb on one of his books: 'His first book, *The Fourth Dimension* (1909) placed him at the age of thirty-one in the front rank of important writers on abstract mathematical theory. His second book, *Tertium Organum*, written in 1912, created a sensation in America on its publication in 1922.'

I have been unable to obtain *The Fourth Dimension*. It would have been most interesting to compare it with his later works. Perhaps one may also label Buckminster Fuller a latter-day mystic. His geodesic domes are not a new idea in the round, as it were. If any of the Platonic solids we have considered are visualized with a sphere passing through their vertices, and rays from the centre of the sphere to the edges trace arcs of great circles on the surface of the sphere, these arcs on the upper half of the sphere look like a geodesic dome. But Buckminster Fuller says that his domes are more cunningly devised.

Before we leave the Platonic solids, we must point out that each of them yields a group, if we consider those symmetries of space which map a given Platonic solid onto itself, the symmetries again being distance-preserving. And having discussed circular cones a number of times, we must draw attention to other surfaces in three dimensions which are *generated by straight lines*.

The cylinder is an obvious example, and it may be regarded as a special case of the cone, when the vertex of the cone is a long way off. Now suppose that we have two equal circles of wire, and construct a cylinder with threads, joining corresponding points on the two circles, which are held parallel to each other (Figure 123a). If we now hold one circle still, and gradually rotate the other, keeping the threads taut, we obtain a new type of surface generated by lines, and a plane section of this surface suggests a hyperbola, with its two branches. If we look down on the surface from above, the section we see is an ellipse, when originally, before the twist, it was a circle. This surface is called *a hyperboloid of one sheet*, and although our threads do not show it, a *second system* of lines lies on the surface, cutting every one of the threads used to indicate the first system (Figure 123b). Any two lines in either system are mutually *skew*, that is they are not parallel and

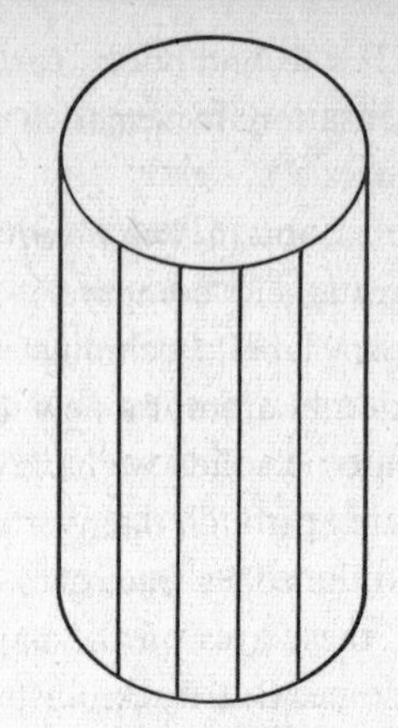

Figure 123a

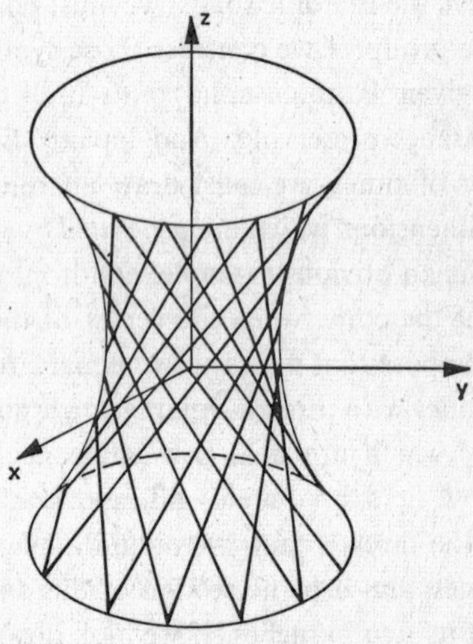

Figure 123b

they do not intersect. But every line of one system meets every line of the other. Some models do show both systems of lines, but they are rather hard to make. A model which shows both systems of lines on the surface is that of the *hyperbolic paraboloid*, and is quite easy to construct.

Cut a square *ABCD* out of cardboard, and divide each of the edges into six equal divisions, boring holes along each edge for threads. Score the cardboard lightly along the diagonal *AC*, so that it can bend without tearing. Fasten threads from the divisions of *AB* to those of *CD*, and threads of another colour from the divisions of *BC* to those of *AD*, and weight the threads with small pieces of lead, used by fishermen, so that the threads are pulled tight by the weights when the distance between the holes lessens. Now fold the cardboard gently along the diagonal *AC*, allowing the weights to function, and the two systems of generators on a hyperbolic paraboloid can be seen (Figure 124a). Each system is

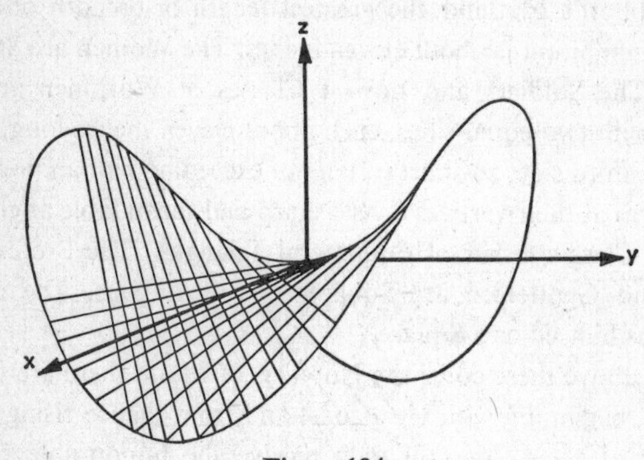

Figure 124a

parallel to a plane. This surface is also a good example for the exhibition of *saddle-points* (Figure 124b). As the reader can imagine, an enormous field of activity opens up before us the moment we begin to discuss surfaces. But we shall now return to two dimensions.

We talked about living in two dimensions, Flatland, when we discussed a model of a non-Euclidean geometry (p. 162). But our survey was a rather restricted one, the universe of the inhabitants being the inside of a circle. In the late nineteenth century Edwin A. Abbott wrote *Flatland, A Romance of Many Dimensions.*

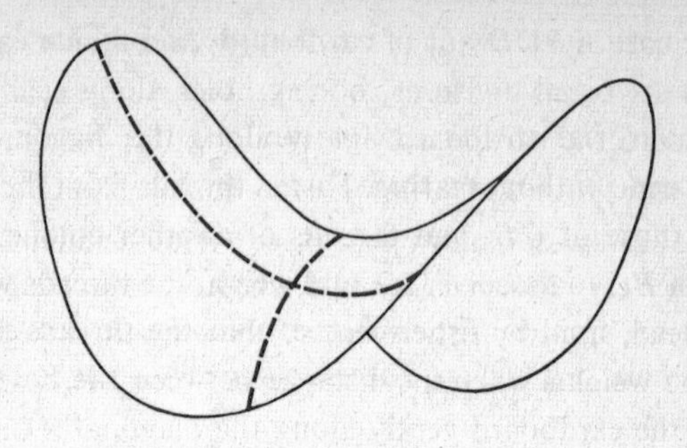

Figure 124b

In Abbott's Flatland, the greatest length or breadth of a full-grown inhabitant is about eleven inches. The Women are Straight Lines. The Soldiers and Lowest Classes of Workmen are Triangles with two equal sides, each about eleven inches long, and a base, or third side, so short (often not exceeding half an inch) that they form at their vertices a very sharp and formidable angle. The Middle Class consists of Equilateral Triangles. The Professional Men and Gentlemen are Squares and Pentagons. The author describes himself as a Square.

Next above these come the Nobility, of whom there are several degrees, beginning with Hexagons, and from thence rising in the number of their sides till they receive the honourable title of Polygonal, or many-sided.

Finally, when the number of sides becomes so very numerous, and the sides themselves so small that the figure cannot be distinguished from a circle, the many-sided one is included in the Circular or Priestly order, and this is the highest class of all.

In Abbott's Flatland angles can be measured. In fact the whole of this delightful story depends on angles. Since, in the plane, all that the inhabitants see of each other is a segment of a line, approaching anyone with a sharp angle can be very dangerous, since one can be pierced. Women especially, being as sharp as needles, and able to make themselves practically invisible by turning at right angles to the person approaching, are feared, and subject to strict laws.

Every house has an entrance on the Eastern side, for the use of Females only, and the law reads:

No Female shall walk in any public place without continually keeping up her Peace-cry, under penalty of death. Any Female, duly certified to be suffering from St Vitus's Dance, chronic cold accompanied by violent sneezing, or any disease necessitating involuntary motions, shall be instantly destroyed.

Since Females are dangerous if they attend public gatherings, some parts of Flatland have an additional law, forbidding Females, under penalty of death, from walking or standing in any public place without moving their backs constantly from left to right to indicate their presence to those behind them.

The writer remarks that whenever the temper of Women is exasperated by confinement at home, or hampering regulations abroad, they are apt to vent their spleen upon their husbands and children, and in the less temperate climates the whole male population of a village has sometimes been destroyed in one or two hours of simultaneous female outbreak.

Recognition of each other by the inhabitants of Flatland is by voice and feel. Long practice and training, begun in the schools and continued in the experience of daily life enables them to discriminate at once by the sense of touch between the angles of an equilateral Triangle, Square and Pentagon, and, of course, the brainless vertex of an acute-angled Isosceles is obvious to the dullest touch. But even a Master of Arts has been known to confuse a ten-sided with a twelve-sided Polygon, and to decide promptly and unhesitatingly between a twenty-sided and twenty-four sided member of the Aristocracy would tax even a Doctor of Science. Because of the importance of angles in Flatland, and the feeling of angles, introductions among the higher classes begin: 'Permit me to ask you to feel and be felt by my friend Mr So-and-so.'

The crisis in the story is occasioned by the sudden appearance of a Circle in the house of the narrator. His grandson has just asked some awkward questions which seem to imply the existence of another dimension, an unthinkable and therefore absurd concept, and has been sent to bed in disgrace. 'The boy is a fool!', says the narrator. At once there comes a distinctly audible reply, 'The boy is not a fool!', and a Figure appears. At first both grandfather and grandmother think it is a Woman, but they soon feel that it is a Circle, and moreover one which changes its size!

The variable Circle asks the Woman to retire, and as her Peace-cry dies away, proceeds to explain that he is a Sphere in three dimensions, and that all that can be visible of him in Flatland is a section, which is a variable circle.

The rest of the story you should read for yourselves. Abbott's book is a worthy companion to Swift's *Gulliver's Travels* on any bookshelf. There have been subsequent imitations, not, I think, very successful ones.

A detailed experience in two dimensions leads quite naturally to a consideration of higher dimensions. Inhabitants of a plane cannot imagine a line at right angles to the plane. When discussing Descartes' introduction of co-ordinates into geometry (p. 176) we considered a set of mutually orthogonal axes for three-dimensional space, passing through a point O, with the co-ordinates (x, y, z) of a point in space being given by the perpendicular distances of P from the planes OYZ, OZX and OXY respectively.

Why may we not assume the existence of a line OT perpendicular to OX, to OY and to OZ? If we do, we consider points with four co-ordinates (x, y, z, t), and we say that we are discussing a space of *four dimensions*.

This suggestion, when it was first made, was deeply resented, as being against the facts of human experience, and even as being anti-religious, but the idea of *time* as a dimension, in a vague sense, is a very old one, and mathematicians found that the problems they were becoming interested in contained more than three varying quantities (x, y, z).

If four varying quantities (x, y, z, t) can be discussed, the first three referring to space coordinates, the fourth t to a time co-ordinate, there is nothing to prevent us discussing n-dimensional space, for any value of n, and, in fact, nowadays undergraduate courses discuss space of n dimensions.

Einstein's theories of relativity gave much prominence to the idea of time being a dimension, and the four-dimensional world of space-time is a fairly familiar notion, which does not mean that everyone understands Einstein's theories.

But many writers have dallied with the idea of a four-dimensional universe, of which our own universe is merely an instantaneous cross-section, and we saw that Abbott in his *Flatland* illustrated the impact of a three-dimensional solid on his two-dimensional people. We shall therefore spend a little time in talking about four dimensions, and some of its special properties, hoping that the incredulity and anger of the inhabitants of Flatland will not be shared by the three-dimensional readers of this book, when the space they live in is enlarged by one dimension.

Television, of course, has seized on the ideas of science-fiction writers (including ghost-story writers) about a mysterious door through which one can pass into another world. Sometimes the door is a mirror-frame, for variety, the image in the mirror changing to reveal the depths beyond. This idea is based on one of the Propositions of Incidence for a space of four dimensions: *two three-dimensional spaces in a space of four dimensions intersect in a plane*. Hence, passing through a door, that is through a plane, could enable one to enter a *new* three-dimensional space.

In a plane, the Propositions of Incidence declare that a unique line can be drawn to connect two distinct points, and two distinct lines intersect in at most one point. If they do not intersect at all, they are said to be parallel.

In three dimensions, the space we inhabit, there are points, lines and planes. A line is *still uniquely determined by two points*, and a plane is uniquely determined if it has to contain three distinct points which do not lie on a line.

To see this, draw any plane through the line which joins two of the points, and rotate the plane about the line until it passes through the third given point. If the plane through the three given points were not unique, we should have two distinct planes intersecting in three *non-collinear* points, which is absurd! Two distinct planes, if they intersect in a point also intersect in a line (Figure

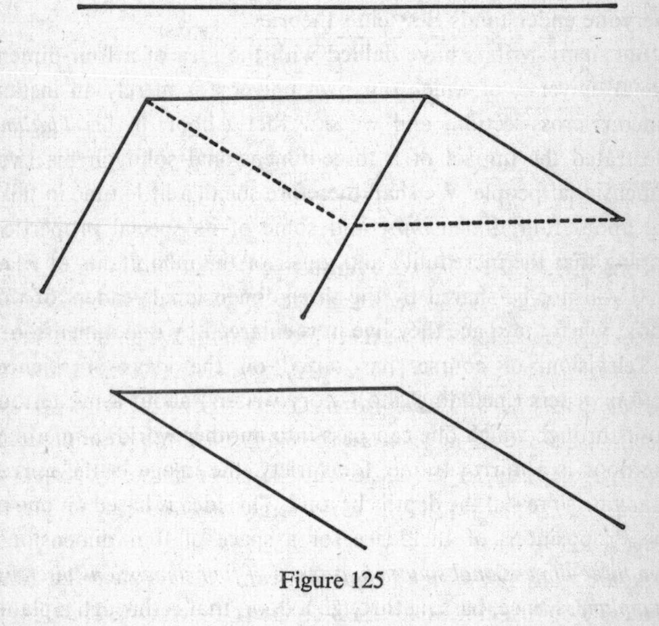

Figure 125

125). A line may be parallel to a plane, and never intersect it, and two planes may be parallel and never intersect.

So we have the Propositions of Incidence in space of three dimensions. Now consider the impression made on the inhabitants of a plane by an object moving along a line which only meets the plane in the point P (Figure 126). The inhabitants of the plane are

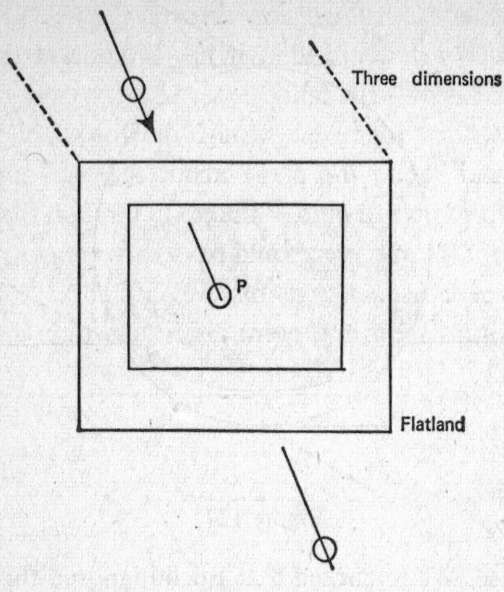

Figure 126

unconscious of the object moving along the line until it is at P. At this instant they see the object, and if it keeps on moving it then vanishes. We already remarked the effect of a sphere in three dimensions passing through a plane (p. 274). At first the inhabitants of Flatland see nothing, then a point appears, at the instant when the sphere *touches* the plane, then the point expands into a larger and larger circle, but the circle suddenly ceases to expand, and then contracts to a point, and the rest is silence (Figure 127)!

Apart from these manifestations from outer space, the Flatlanders would become extremely agitated if a triangle ABC were taken up out of the plane, turned over, and then put down again. The relation between the original triangle and the one taken up and laid down again, which we call $A'B'C'$, is that one can be moved to become the mirror image of the other (Figure 128). If a Flatlander felt his way around the triangle ABC in a clockwise sense, he would feel his way around triangle $A'B'C'$ in an anti-

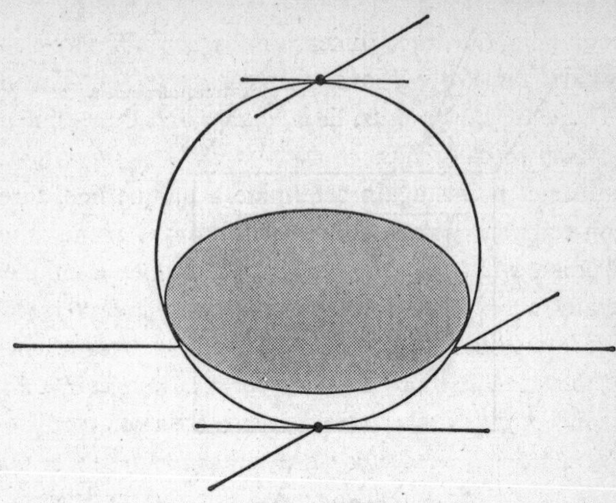

Figure 127

clockwise sense. We remarked that Euclid ignored this difference between congruent triangles in his proof of the Side-Angle-Side Proposition.

To see what kind of tricks an inhabitant of four-dimensional

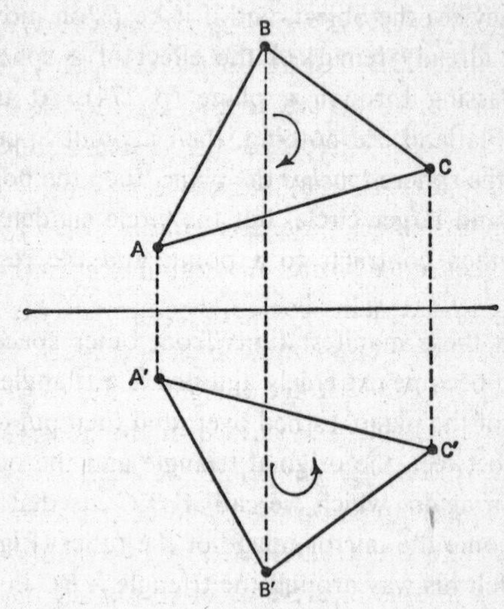

Figure 128

space containing our three-dimensional space can play on us, we first consider the Propositions of Incidence in four-dimensional space. There are now points, lines, planes and three-dimensional spaces, which we call *solids.*

Two distinct points again determine a unique line, three distinct non-collinear points determine a unique plane, and four distinct points which are non-coplanar determine a unique solid.

Generally, a given line and a given plane do *not* intersect, but a given line intersects a solid in a point. Two given planes meet in a point. If they meet in more than one point they meet in a line, and lie in a solid. A given plane meets a given solid in a line, or lies in the solid, and two given solids which are distinct meet in a plane.

There is a simple formula which connects the dimensions of the pairs of spaces we have been considering with the dimension of their intersection, and this is:

Sum of dimensions of the two spaces =

4 + *dimension of intersection.*

If this formula gives a dimension of intersection less than zero (the dimension of a point is zero), the two spaces do not intersect. For example, a general line (dimension 1) and a general plane (dimension 2) do not intersect. If they do, they lie in a space of *three* dimensions. We verify that the formula shows that two solids (both have dimension 3) intersect in a plane.

It is through this plane of intersection that we may pass to enter a new three-dimensional world. We examine the impression made upon someone who remains behind, when an entry is made into a distinct solid by means of an interconnecting door.

In Figure 129 let PQ be a line at right angles to the door, with the point Q just in the plane surface of the door. The moment the point Q moves into the new solid, *the whole line* PQ *vanishes,* leaving only a point trace on the plane of the door. This is because a line is completely determined by two points on it, and if two points of a line lie in a plane, or in a solid, the whole line lies in that plane, or solid.

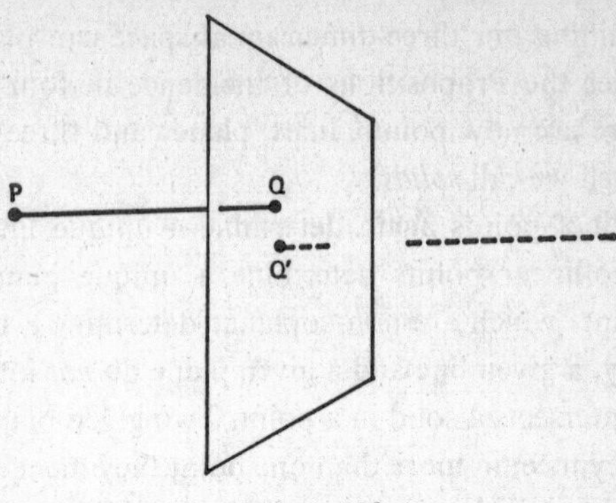

Figure 129

If Q moves into the new solid, and the line intersects the plane of the door in Q', the line QQ' lies entirely within the new solid, and therefore vanishes from our sight, assuming that we choose to remain behind.

If this seems unbelievable, recollect what things look like to an inhabitant of Flatland if a line PQ in Flatland is lifted even an infinitesimal distance out of the plane. The line vanishes, leaving just the trace where the new line meets the plane, the point Q'.

Returning to our magic door, suppose that we have two closed boxes, one inside the other, and wish to free the inner one without opening the outer one. Move them so that the outer box has one face resting in the plane of the magic door, then, without disturbing the inner box, move the outer box gently so that the face in the plane of the magic door moves a slight distance into the new solid. The outer box immediately *vanishes*, leaving just a trace of one side in the plane of the door, and the inner box is free, and can be removed, since our very small displacement of the outer box has not disturbed the inner box from its position in our three-dimensional space.

It is a disturbing thought that an eye in four-dimensional space can see what goes on inside a windowless room in our space. Let E be the eye, and P any object in the room. The line PE joining the object to the eye does not intersect any wall, ceiling or floor of the room. For suppose that it does so at the point Q, which would prevent the rays of light from P reaching the eye at E. Then since the points P and Q both lie in our three-dimensional space, the *whole line* PQ does so, and the point E, which is on the line, also does so. But this is a contradiction, since E is assumed to lie outside our space of three dimensions. Hence all points in the room are visible (Figure 130).

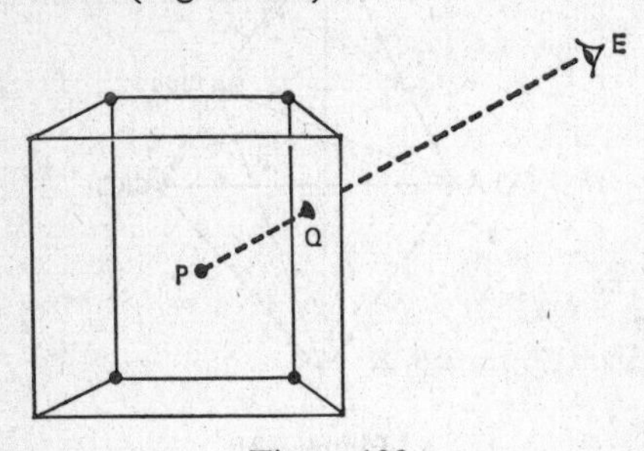

Figure 130

This property of four-dimensional space is mentioned by H. G. Wells in *The Plattner Story*. A young teacher of languages, Gottfried Plattner, also teaches chemistry. In the course of an experiment he blows himself into four dimensions, and witnesses a murder taking place inside a shuttered room in our space. Eventually, after an absence of nine days, Gottfried blows himself back again into our space, landing on his headmaster's back while the latter is browsing in a strawberry patch! It is remarked that the unlucky young man on his return is a mirror image of what he was before he left, his heart, in particular, now beating on the right side. Hands are mirror images anyway, so his hands look the same, but his sense of right and left is confused.

We saw that a triangle can have its *orientation* changed (the sense of traversing the vertices A, B and C), by being lifted up out of its plane, turned over and put back. The new triangle can be moved so as to be a mirror image of the original one.

The two tetrahedra in Figure 131, $ABCD$ and $A'B'C'D'$ have all their corresponding sides equal, but to an observer at D the orientation of triangle ABC differs from that of triangle $A'B'C'$ to an observer at D'. One tetrad can be obtained from the other by a removal into a space of four dimensions, and a replacement

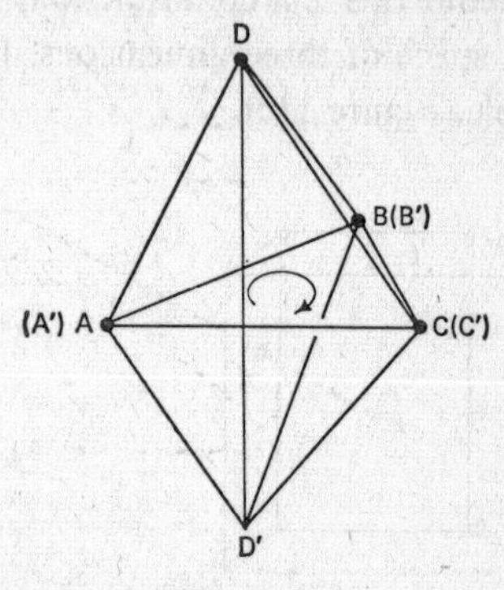

Figure 131

in our space. We note that here also one can be obtained from the other as a mirror image in the plane ABC.

For Gottfried Plattner the mirror is a plane bisecting him at right angles. Wells has another story, his first, and perhaps his greatest, *The Time Machine*, in which the idea of time as a dimension is very clearly explained. And, of course, *The Door in the Wall* is also about an adventure in another space. Recent television programmes indicate that few people nowadays are as familiar with the properties of four-dimensional space as H. G. Wells was. In these programmes the person who walks through the magic door or mirror into a new solid space can still be seen by those in our space, and as we have shown already, this would not be possible.

As a final comment on the possibilities of four-dimensional space, it is interesting to speculate that the fate of the crew of the *Marie Celeste* may be bound up with their vanishing into another space. This ship was found in a condition suggesting that the crew had suddenly and most inexplicably vanished. No trace of the crew has ever been discovered. Where could they have gone?

The reader may feel that, all the same, there cannot be such a thing as space of four dimensions, even as an abstract construct. We therefore end this chapter with a gradual development of the notion of a four-dimensional cube. Although it is not essential for our purpose, the use of co-ordinates will be helpful.

We begin with a two-dimensional cube, that is, a square (Figure 132) with one vertex at the origin of co-ordinates (0, 0), and two

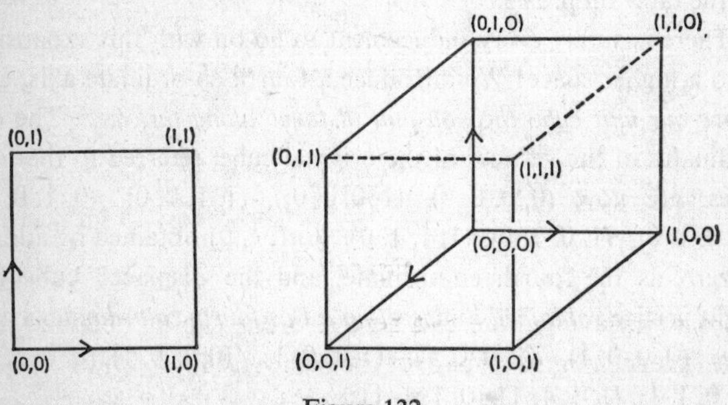

Figure 132

sides along the co-ordinate axes. The co-ordinates of the four vertices of the square are therefore (0, 0), (1, 0), (1, 1) and (0, 1).

This square has four bounding sides. If we introduce a third co-ordinate axis, we can imagine the square moved through unit distance along this axis. It then traces out a *three*-dimensional cube, our ordinary and every-day cube.

The co-ordinates of the vertices of this cube are:

(0, 0, 0), (1, 0, 0), (1, 1, 0), (0, 1, 0)

(which are the co-ordinates of the original square referred to the original axes and the new axis, and obtained from the original co-ordinates by adding a zero as the third co-ordinate), and further:

(0, 0, 1), (1, 0, 1), (1, 1, 1) and (0, 1, 1),

these being the four new vertices introduced by moving the square a unit distance along the new axis. All that we have to do to obtain these *new* vertices is to *put a 1 as the third co-ordinate* of the original vertices.

The three-dimensional cube has eight vertices, and is bounded by the original square and its final position, and the additional four squares swept out by the sides of the original square as they move through unit distance. The three-dimensional cube therefore has eight vertices and six bounding faces, which corresponds to the table on p. 263.

There is surely every inducement to go on with this expansion into a higher space! We introduce a fourth co-ordinate axis, and *move our unit cube through unit distance along this axis.* The co-ordinates of the vertices of the original cube, referred to the four axes, are now (0, 0, 0, 0), (1, 0, 0, 0,), (1, 1, 0, 0), (0, 1, 0, 0), (0, 0, 1, 0), (1, 0, 1, 0), (1, 1, 1, 0), (0, 1, 1, 0) obtained by adding a zero as the fourth co-ordinate, and the displaced cube has eight vertices *obtained by inserting a 1 as fourth co-ordinate.* These are: (0, 0, 0, 1), (1, 0, 0, 1), (1, 1, 0, 1), (0, 1, 0, 1), (0, 0, 1, 1), (1, 0, 1, 1), (1, 1, 1, 1), (0, 1, 1, 1).

We have now constructed a four-dimensional cube. It has sixteen vertices, and besides the original and final positions of the cube, which are three-dimensional bounding solids, each face of the original three-dimensional cube sweeps out a three-dimensional bounding cube as it moves, so that there are an extra six bounding three-dimensional cubes, and the total number of three-dimensional bounding cubes is eight.

It is possible to draw a net, a three-dimensional cut-out for this cube in space of three dimensions, which if stuck together in the right way would give our four-dimensional cube, if we could move into four-dimensional space, but we shall not do so. With the co-ordinates of the sixteen vertices displayed, we know that it exists!

Exercises

1. Can you devise a method by which the inhabitants of Flatland could recognize an acute angle approaching, so that the Women would have an easier life, not always trilling their Peace-cry?

2. Draw a net in two dimensions of an ordinary cube, and then construct the cube. Can you sketch a four-dimensional cube as a net in three dimensions?

3. What do you think happened to H. G. Wells's Time-Traveller as he sped forward through time, and as he sped backward? Does H. G. Wells consider the possible biological changes?

4. Do you have any definite objections to the possibility of our universe being a cross-section of a four-dimensional continuum?

5. In *The Gentle Art of Mathematics* (p. 79 of the E.U.P. edition) the reader is shown how to take off a waistcoat without taking his arms out of his jacket first. Can you devise an easier method, by using four dimensions?

6. Show how the knot illustrated below may be untied, without being cut, by use of space of four dimensions. (Hint: move a piece of the knot into another solid, the remainder being held in our space. What happens?)

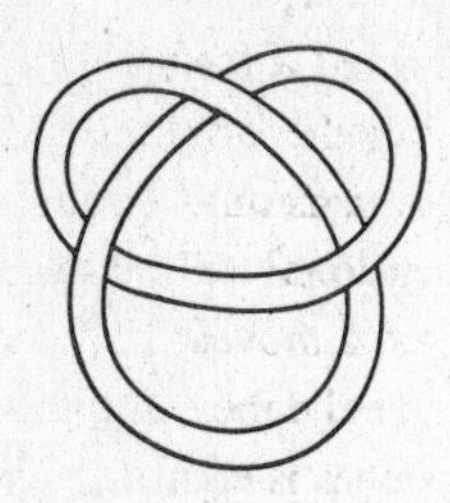

BIBLIOGRAPHY

ABBOTT, E. A., *Flatland* (Dover, 1952)

COOK, SIR THEODORE, *The Curves of Life* (London, 1914)

DICK, OLIVER LAWSON, *Aubrey's Brief Lives* (Secker & Warburg, 1949 and Penguin Books, 1972)

GOMBRICH, ERNST H., *Art and Illusion* (Random House, 1960)

HEATH, SIR T. L., *The Thirteen Books of Euclid's Elements* (Cambridge University Press, 1928)

Apollonius of Perga, The Treatise on the Conic Sections (Cambridge University Press, 1896)

LOCKWOOD, E. H., *A Book of Curves* (Cambridge University Press, 1961)

MORGAN, MORRIS HICKY, *Vitruvius, The Ten Books on Architecture* (Dover, 1960)

MORROW, GLEN R. (Ed.), *Proclus' Commentary on the First Book of Euclid's Elements* (Princeton University Press, 1970)

PANOFSKY, ERWIN, *The Life and Art of Albrecht Dürer* (Princeton University Press, 1943)

PEDOE, DAN, *The Gentle Art of Mathematics* (English University Press, 1965, and Dover, 1973)

PERRAULT, CLAUDE, *Vitruvius, The Ten Books on Architecture* (Cognard, Paris, 1684)

PIRENNE, M. H., *Optics, Painting and Photography* (Cambridge University Press, 1970)

SCHOLFIELD, P. H., *The Theory of Proportion in Architecture* (Cambridge University Press, 1958)

THOMPSON, D'ARCY W., *On Growth and Form* (Cambridge University Press, 1952)

WENNINGER, MAGNUS J., *Polyhedron Models* (Cambridge University Press, 1971)

WEYL, HERMANN, *Symmetry* (Princeton University Press, 1952)

INDEX